T0211925

Reinforcement Learning-Enabled Intelligent Energy Management for Hybrid Electric Vehicles

Synthesis Lectures on Advances in Automotive Technology

Editor
Amir Khajepour, *University of Waterloo*

The automotive industry has entered a transformational period that will see an unprecedented evolution in the technological capabilities of vehicles. Significant advances in new manufacturing techniques, low-cost sensors, high processing power, and ubiquitous real-time access to information mean that vehicles are rapidly changing and growing in complexity. These new technologies—including the inevitable evolution toward autonomous vehicles—will ultimately deliver substantial benefits to drivers, passengers, and the environment. Synthesis Lectures on Advances in Automotive Technology Series is intended to introduce such new transformational technologies in the automotive industry to its readers.

Reinforcement Learning-Enabled Intelligent Energy Management for Hybrid Electric Vehicles
Teng Liu
2019

Narrow Tilting Vehicles: Mechanism, Dynamics, and Control
Chen Tang and Amir Khajepour
2019

Dynamic Stability and Control of Tripped and Untripped Vehicle Rollover
Zhilin Jin, Bin Li, and Jungxuan Li
2019

Real-Time Road Profile Identification and Monitoring: Theory and Application
Yechen Qin, Hong Wang, Yanjun Huang, and Xiaolin Tang
2018

Noise and Torsional Vibration Analysis of Hybrid Vehicles
Xiaolin Tang, Yanjun Huang, Hong Wang, and Yechen Qin
2018

Smart Charging and Anti-Idling Systems
Yanjun Huang, Soheil Mohagheghi Fard, Milad Khazraee, Hong Wang, and Amir Khajepour
2018

Design and Avanced Robust Chassis Dynamics Control for X-by-Wire Unmanned Ground Vehicle
Jun Ni, Jibin Hu, and Changle Xiang
2018

Electrification of Heavy-Duty Construction Vehicles
Hong Wang, Yanjun Huang, Amir Khajepour, and Chuan Hu
2017

Vehicle Suspension System Technology and Design
Avesta Goodarzi and Amir Khajepour
2017

© Springer Nature Switzerland AG 2022
Reprint of original edition © Morgan & Claypool 2019

All rights reserved. No part of this publication may be reproduced, stored in a retrieval system, or transmitted in any form or by any means—electronic, mechanical, photocopy, recording, or any other except for brief quotations in printed reviews, without the prior permission of the publisher.

Reinforcement Learning-Enabled Intelligent Energy Management for Hybrid Electric Vehicles
Teng Liu

ISBN: 978-3-031-00375-2   paperback
ISBN: 978-3-031-01503-8   ebook
ISBN: 978-3-031-00008-9   hardcover

DOI 10.1007/978-3-031-01503-8

A Publication in the Springer series
SYNTHESIS LECTURES ON ADVANCES IN AUTOMOTIVE TECHNOLOGY

Lecture #9
Series Editor: Amir Khajepour, *University of Waterloo*
Series ISSN
Print 2576-8107   Electronic 2576-8131

Reinforcement Learning-Enabled Intelligent Energy Management for Hybrid Electric Vehicles

Teng Liu
University of Waterloo

*SYNTHESIS LECTURES ON ADVANCES IN AUTOMOTIVE TECHNOLOGY
#9*

ABSTRACT

Powertrain electrification, fuel decarburization, and energy diversification are techniques that are spreading all over the world, leading to cleaner and more efficient vehicles. Hybrid electric vehicles (HEVs) are considered a promising technology today to address growing air pollution and energy deprivation. To realize these gains and still maintain good performance, it is critical for HEVs to have sophisticated energy management systems. Supervised by such a system, HEVs could operate in different modes, such as full electric mode and power split mode. Hence, researching and constructing advanced energy management strategies (EMSs) is important for HEVs performance. There are a few books about rule- and optimization-based approaches for formulating energy management systems. Most of them concern traditional techniques and their efforts focus on searching for optimal control policies offline. There is still much room to introduce learning-enabled energy management systems founded in artificial intelligence and their real-time evaluation and application.

In this book, a series hybrid electric vehicle was considered as the powertrain model, to describe and analyze a reinforcement learning (RL)-enabled intelligent energy management system. The proposed system can not only integrate predictive road information but also achieve online learning and updating. Detailed powertrain modeling, predictive algorithms, and online updating technology are involved, and evaluation and verification of the presented energy management system is conducted and executed.

KEYWORDS

energy management, hybrid electric vehicles, reinforcement learning, deep learning, intelligent transportation information, real-time updating, velocity and power prediction, optimality and adaptability

Contents

Preface

Electrified powertrains are encouraged as potential solutions to environmental concerns, desire for mobility, and safety concerns. Hybrid vehicles have been on the scene since the successful development of Prius by Toyota and Insight by Honda. Energy management strategies play a significant role in hybrid electric vehicles to achieve the goal of fuel economy improvement and pollution emissions reduction. Thanks to the existence of the internal combustion engine, battery pack, generator, and electric motor, the hybrid powertrain can work in pure electric and hybrid electric mode to accommodate driving conditions.

This book focuses on developing reinforcement learning-enabled energy management strategies for different hybrid powertrains. To the best of my knowledge, there has not been a book that systematically discusses the energy management strategies founded in artificial intelligence. Hence, this book aims to introduce the fundamental elements of reinforcement learning and various applications of reinforcement learning algorithms in the energy management field. The material assembled in this book is an outgrowth of the experience that the author gained while studying as a Ph.D. student in Beijing Institute of Technology, China, and working as a Postdoctoral Fellow of University of Waterloo, Canada.

The text mainly discusses the applications of reinforcement learning in energy management fields, including the optimality of signal algorithm and the mixture of multiple techniques. To apply the proposed system, methods, and framework into real-time applications of hybrid electric vehicles is the author's constant objective. I hope this book will interest graduate students, practitioners, and vehicle engineers in the area of hybrid electric vehicles.

Finally, I would like to express my heartfelt gratitude for the guidance and help of my family, supervisor, and colleagues. I am also grateful to all the editors for their support and patience in the production of this book. This book is also for my younger daughter, who was born on June 25, 2019.

Teng Liu
July 2019

C H A P T E R   1

Introduction

1.1 MOTIVATION

Nowadays, increasing fuel awareness and environmental concerns contribute to the development and production of hybrid vehicles. The propulsion system of these vehicles consists of multiple energy storage sources (ESSs): a lower-capacity ESS to provide power assistance and recover powertrain kinetic energy, and a high-capacity ESS to supply the main power. Generally, the high-capacity ESS is an internal combustion engine (ICE) and the lower-capacity ESS can be electrochemical (batteries or supercapacitors), hydraulic/pneumatic (accumulators), or mechanical (flywheel) [1]. The electrochemical batteries and ICE are the common combination of hybrid vehicles on the road, which power the hybrid electric vehicles (HEVs) [2]. These vehicles have the potential to downsize the ICE, reduce pollutant emissions, and improve energy efficiency through reasonable technologies, and thus they are very popular and welcome by vehicle manufacturers [3].

The power supplied by the energy sources is transmitted to wheels through an electric machine (EM) and a transmission to drive the vehicle. According to the featured functions of the propulsion system, hybrid electric vehicles are divided into three categories, which are micro, mild, and full hybrids [4]. Micro hybrids only contain the start/stop system, which enables the ICE to shut down and restart to reduce air pollution and fuel consumption [5]. Mild hybrid vehicles are equipped with ICE and EM to realize regenerative braking and power assistance, but the powertrain does not have a pure electric mode. Full hybrids not only include the abilities of mild hybrids but also electric launch. More importantly, efficient energy management strategies (EMSs) are necessary for full hybrids to adequately take advantage of the benefits of powertrain hybridization [6]. Besides this classification, plug-in hybrid electric vehicles (PHEVs) and battery electric vehicles (BEVs) are seen as long-term solutions after overcoming critical factors such as insufficient running range, high battery cost, long charging duration, and charging infrastructure deficiency. The evolution process from conventional vehicles to BEVs is displayed in Fig. 1.1. This book focuses on full hybrid electric vehicles.

Energy management is the key technology for an HEV to exploit its advantage in energy saving and emission reduction [7]. It represents the decision of power distribution between multiple ESSs in order to maximize or minimize predefined objectives while meeting several constraints. The defined objectives are expressed as reducing exhaust emissions, delaying battery aging, maintaining vehicle mobility and drivability, and minimizing fuel consumption [8, 9]. The constraints refer to the variables of onboard components, such as the state of charge (SOC)

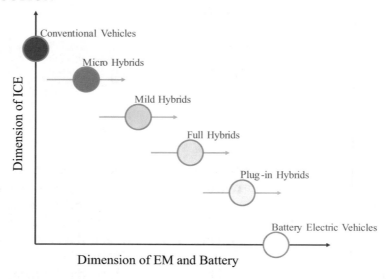

Figure 1.1: Evolution and development of vehicle hybridization.

and state of health (SOH) in the battery, speed and torque limitations in ICE and generator, battery current and power, and dynamic characteristics [10, 11]. Generally, the inputs of the energy management module are power demand, vehicle speed, and current SOC value, and its outputs are the amount of delivered power by the onboard energy sources.

Artificial intelligence (AI) and machine learning (ML) have become powerful measures and study hotspots in many research fields [12–15]. Reinforcement learning (RL) is a representative paradigm of ML to decide appropriate actions by interacting with an environment to maximize the cumulative reward. Many problems can fall into its formulation and be solved by RL, such as AlphaGo [16], robot control [17], elevator scheduling [18], etc. Given RL's merits in calculative efficiency and real-time applications, the book aims to derive the RL-enabled intelligent energy management system for HEVs. The discussed powertrain is a series paradigm, and both current and future driving conditions are considered in this work. This book will hopefully make some contributions to intelligent and real-time energy management strategies for various configurations of HEVs.

1.2   HEV POWERTRAIN

Although the components of HEV powertrain are determined, different powertrain configurations can be constructed. HEV powertrain can be classified into three types: series, parallel, and power split architectures [19]. The schematic diagrams of these constructions are shown in Figs. 1.2, 1.3, and 1.4, respectively. In series powertrain, the electric power from the engine-generator set and battery are combined and transmitted to the EM to drive the wheels. For

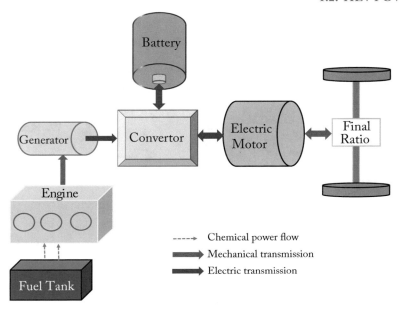

Figure 1.2: Configuration of the series HEV powertrain.

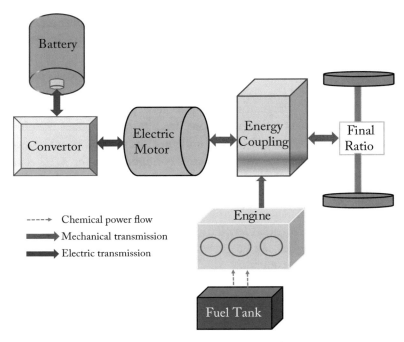

Figure 1.3: General architecture of parallel hybrid electric powertrain.

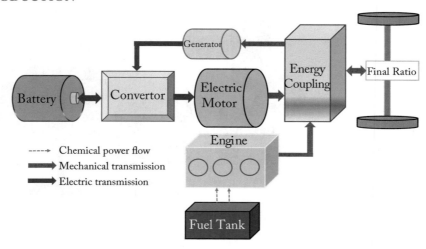

Figure 1.4: Configuration of the power split HEV powertrain.

parallel configuration, the electric power or the mechanical could propel the wheels alone, and they could also be summed with a gear set or a chain to drive the wheels jointly. Generally, a planetary gear set is employed to connect the ICE, EM, and generator in power split architecture. The hybrid vehicle could work in different operation modes to export mechanical or electric power to drive the vehicle. The power from the engine can be decided by an energy management policy to propel the vehicle or charge the battery.

The pros and cons of these three architectures are obvious. The series pattern requires only electric connections between the main power output devices [20]. The disconnection between the ICE and wheels has the advantage of selecting feasible speed and torque for the engine, and thus it can operate at the highest efficiency. However, appropriate sizes should be chosen for the electromechanical energy converters (motor and generator) to support the maximum power demand of the powertrain. Compared with the series configuration, the parallel architecture has a smaller size EM because the engine could drive the wheels alone and not all the mechanical power would flow through the motor. As ICE speed is related to the vehicle speed in this construction, ICE working conditions should be determined with particular constraints. Finally, the power split hybrid powertrain has great freedom in operation criteria of the ESSs (engine and battery). Multiple operation modes can be chosen to accommodate different driving situations, thus improving overall energy efficiency.

1.3   LITERATURE REVIEW

HEVs' energy management strategies are typically divided into two categories, i.e., rule-based and optimization-based. Rule-based energy management policies are always expressed as a set of predefined thresholds with respect to the control actions and are extracted from human engineering experiences or the obtained optimal control policies. Optimization-based ones are usually derived by using various optimal control algorithms, wherein an optimal control problem is built with control objective and physical constraints. Since HEVs began emerging on public roads over the last few years, many attempts have been made to formulate the energy management system of HEVs. Figure 1.5 depicts the classification of existing energy management strategies.

Figure 1.5: Classification of energy management strategies for HEVs.

1.3.1 REVIEW LITERATURE

Typically, the current literature on EMSs of hybrid electric vehicles is of two different types: the first is comprehensive review papers in terms of different points of view, and the other is regular articles seeking to elaborate advanced algorithms and techniques in the energy management domain. Many researchers have proposed survey manuscripts according to various perspectives, such as global optimization control methods, connected vehicles view, a combination of energy management and component sizing, specific model predictive control (MPC) evolution for HEVs, etc. For example, Serrao et al. compared three global optimal approaches for energy management in 2011 [21], which are dynamic programming (DP), Pontryagin's minimum principle (PMP), and equivalent consumption minimization strategy (ECMS). These methods-based results served as the benchmark in following years. The authors in [22] and [23] summarized and analyzed the EMSs as two independent topics, MPC-based ones and integrated ones considering power management and component size simultaneously. Owing to the diversity of energy management methods, Refs. [24–27] conducted an extensive review of the existing algorithms for HEVs, wherein the three important powertrain architectures, as well as online and offline techniques, are discussed.

Furthermore, Martinez et al. highlighted the EMSs in the context of connected vehicles and the outlook for future trends of PHEVs in the intelligent transportation system (ITS) [28]. Designing a through-the-road HEV with in-wheel motor has been a popular concept in recent years. The authors in [29] determined the pros and cons of this idea, and also compared its performance with conventional HEVs. Aiming principally at a parallel hybrid electric vehicle, Enang et al. explained the realization process of different control methods on this configuration [30], including the workflow, equations, and parameters. A detailed overview of the review literature for energy management is depicted in Table 1.1.

Table 1.1: Content of current review papers in HEV's energy management

References	Powertrain Architecture	Content Description
[21]	Series HEV	Describes and analyzes DP, PMP, and ECMS
[22]	HEVs and PHEVs	Elaborates MPC-based power management strategies and future study
[23]	All types of HEVs	Discusses factors that affect the performance of EMS and component sizing
[24–27]	All types of HEVs	Classifies EMSs into online and offline types, and explains their pros and cons
[28]	PHEVs in connected environments	Highlights benefits of ITS, traffic information and cloud computing in EMSs
[29]	Through-the-road HEVs	Analyzes the concept of HEVs with in-wheel motor and the related EMSs
[30]	Parallel HEV	Explains the realization process of popular algorithms in the energy management field

1.3.2   ALGORITHM LITERATURE

The hotspot algorithms in the energy management field are mainly classified into two categories, online and offline. The offline EMSs are usually formulated based on optimal control theory, which indicates that they require the particular driving cycle information in advance. They often carry heavy computation burden, and thus they are derived offline; however, they can be treated as a baseline to validate the effectiveness and correctness of other approaches. The relevant offline algorithms are DP, PMP, ECMS, genetic algorithm (GA), bee algorithm (BA), simulated annealing (SA), particle swarm optimization (PSO), convex programming (CP), and game theory (GT), among others.

PSO was first used in [31] to search the global power split controls. The amount of time used and sub-optimal snare prevent wide-range application. Different attempts of ECMS are executed in [32] and [33] and are called adaptive ECMS (A-ECMS) and telemetric ECMS (T-ECMS), in which the tuning rule of co-states is the critical point in this method. DP is one of the most common algorithms for global optimal controls [34–36], and its related results are often utilized to evaluate other novel techniques. Recently, Refs. [37] and [38] tried to extract the DP-correlated control criteria for real-time applications, but the dependency of driving cycles cannot be overcome easily.

By defining the appropriate initial population and tuning parameters, GA has the ability to carry out a global optimal search [39]. SA and BA have a faster convergence rate than GA [40], [41], but they may trap in local optimum due to the enormous state space. Moreover, the authors in [42] and [43] applied CP to solve energy management of fuel cell HEV and considered engine start and gearshift cost. The convex modeling of the powertrain is the pivotal point in this approach and it is not easily extended to the complicated powertrain. GT is suitable to manage the interaction between two agents, and thus Refs. [44] and [45] employed this method to handle the energy management and charging strategies for PHEVs. Up to this point, the offline algorithms could be leveraged in different powertrain and problems; however, the necessity of prior driving condition information stop them from being applied in real-world environments.

The online EMSs are further divided into two categories according to time evolution: rule-based and instantaneous control. Rule-based algorithms usually depend on human experience or engineering knowledge, and they are represented as certain criteria of some arguments, such as engine torque, SOC in battery, and generator speed. To enhance the performance of rule-based EMSs in fuel economy and pollutant emissions, many researchers have proposed some advanced methods to generate rules, such as fuzzy-logic rule [46], power follower policy [47], and on/off strategy [48]. Unfortunately, there is a lot of space to fill when comparing with the global optimum.

As alternatives, instantaneous control algorithms could achieve better performance while obtaining online control implementation. The related algorithms are MPC, stochastic dynamic programming (SDP), sliding mode control (SMC), neural network (NN), extremum seeking control (ESC), and reinforcement learning (RL), among others. For example, many trials have been conducted in MPC, such as stochastic MPC [49], nonlinear MPC [50] and linear varying-time MPC [51]. To effectively apply MPC in energy management problems, high prediction and modeling accuracy are necessary. The authors in [52] considered the battery and super-capacitor in fully active HEV's energy management, and a sliding-mode controller is built to control their currents to reference values. As the hybrid city bus has a regular route, an NN-based network is designed to train the length ratio and achieve online control [53]. The related results can serve as a sub-optimal strategy.

Learning-based EMSs founded in artificial intelligence are more and more welcome in recent years. For example, the authors in [54] and [55] examine the optimality and adaptability of RL-based EMSs via comparison with the DP algorithm. The online RL-based power management policies integrated current and predictive driving cycle information and are introduced in [56–58]. To fuse huge driving data to adapt to various driving situations, deep learning (DL) and RL are combined to derive online EMSs [59–62]. Specifically, Ref. [63] adopted a deep neural network (DNN) to train the action value function in the RL framework and employed the Q-learning algorithm to compute the online controls. As a result, the obtained controls are free of the powertrain modeling and driving cycles.

The authors in [64] constructed deep reinforcement learning (DRL)—enabled power split controls based on stochastic driver models EM, and thus the results showed great potential to improve intelligence and adaptability. Furthermore, the energy efficiency of PHEV can be improved in ITS by sharing real-time traffic conditions (driving cycles) with wireless communication, a global positioning system (GPS), or geographical information systems (GIS) [65, 66]. A comprehensive overview of the different kinds of algorithms in the energy management field is displayed in Table 1.2.

1.4   SUMMARY

In the analysis of hybrid powertrain and energy management system, EMSs are treated as a significant and potential technology to improve fuel economy and reduce global warming. The variety of hybrid powertrain and uncertainty of driving environments increase the difficulty in developing energy management policies. To address these problems, many attempts have been made to study and apply different methodologies for energy management. Motivated by the potential of artificial intelligence and machine learning for enormous data processing, deep learning and reinforcement learning are treated as powerful and useful tools to derive real-time and model-free energy management system. The relevant EMS could be adaptive to different driving styles, driving situations, and driving intention. Therefore, this book aims to discuss the RL-based energy management system by considering online and future driving information.

The remaining content of this book is arranged as follows. Chapter 2 provides an overview of control-oriented modeling approaches for hybrid electric powertrain and reinforcement learning formulation for energy management problem. Chapter 3 describes the online updating algorithm to integrate current driving information and three different methods for future driving conditions prediction. Chapter 4 evaluates the proposed intelligent RL-enabled energy management system by comparing with two benchmark methods: dynamic programming-based and stochastic dynamic programming-based controller. Finally, Chapter 5 is the conclusion of this book.

Table 1.2: Comparative analysis of main algorithms in the energy management field

Algorithms	References	Categories	Description of Characteristics
DP	[34-36], [37, 38], [4], [16]	Offline EMS	Global optimality, dependency of driving cycles
PMP, ECMS	[32, 33], [4]	Offline/Online EMS	Key point is tuning co-states to accommodate driving conditions
GA	[39], [9]	Offline/Online EMS	Global optimality, defining initial population and parameters are core
BA	[41]	Offline EMS	Global optimality, better convergence rate and worse results than GA
SA	[40]	Offline EMS	Time-consuming, fluctuation of performance is large
PSO	[31]	Offline EMS	Suitable for multi-goals with random search strategy
CP	[42, 43]	Offline EMS	High requirement in convex modeling, less computation burden
GT	[44, 45]	Offline EMS	Time-consuming, high dependency on modeling construction
Fuzzy logic rules	[46]	Online EMS, rule-based	Online achievement, performance far from global optimum
Power follower	[47]	Online EMS, rule-based	Online achievement, requirement in special driving situations
On/off strategy	[48]	Online EMS, rule-based	Online achievement, worse than optimization-based results
SDP	[49-51]	Online EMS, instant.*	Data dependency, time-consuming, online EMS
SMC	[52]	Online EMS, instant.	Less applications, reference trajectories are necessary
NN	[53], [63]	Online EMS, instant.	Data dependency, high performance, tuning of parameters
RL	[54, 55], [56-58]	Online EMS, instant.	Real-time implementation, requirement in high-quality controller
Deep RL	[59-62]	Online EMS, instant.	Founded in artificial intelligence, popular methods, real-time EMS

* Instant indicates instantaneous power split controls

CHAPTER 2

Powertrain Modeling and Reinforcement Learning

2.1 CONTROL-ORIENTED MODELING

Energy management strategy aims to distribute power requirement into multiple onboard ESSs while minimizing/maximizing the predefined cost function (optimization objective). To this end, the detailed modeling of the powertrain components need to be created. Two perspectives are often leveraged to establish modeling for energy management problem: simulator-oriented and control-oriented. The former is suitable for testing the performance of existing energy management policy and the latter is appropriate to derive and develop an energy management strategy. The essential goal of these two models is generating an accurate estimation of fuel consumption in ICE and SOC value in the battery.

This chapter constructs a concise overview of control-oriented modeling for hybrid electric powertrain, including the ICE, battery, EM, generator, and transmission. A series hybrid powertrain is mainly considered as the modeling target; this book focuses on discussing intelligent smart methods to solve energy management problem. The essential parameters of this series powertrain are summarized in Table 2.1. However, the mentioned techniques and algorithms could be expanded to parallel and power split powertrain. The influences of temperature and electric accessories are not addressed in this book.

2.1.1 TRANSMISSION MODELING

To propel the wheels of HEV, the traction force F_{tra} generated by the powertrain should equal the combination of various resistances as follows [56]:

$$F_{tra} = F_{ine} + F_{rol} + F_{aer} + F_{gra}, \qquad (2.1)$$

where F_{ine}, F_{rol}, F_{aer}, and F_{gra} are the inertial force, rolling resistance, aerodynamic resistance and road grade resistance, respectively. These resistances are usually expressed by the parameters

Table 2.1: Architecture parameters of the studied series HEV powertrain

Definition	Factor	Unit
Electromotive force factor K_e	1.65	Vsrad^{-2}
Electromotive force factor K_x	0.00037	NmA^{-2}
Minimum value of SOC SOC_{min}	0.2	/
Maximum value of SOC SOC_{max}	0.9	/
Rated capacity of battery Q_c	50	Ah
Vehicle frontal area A	5.4	m^2
Aerodynamic drag coefficient C_d	0.9	/
Air density ρ_{air}	1.2258	Ns^{2}m^{-4}
Vehicle mass M_v	15200	kg
Vehicle tread B	2.55	m
Final transmission ratio i_0	13.2	/
Rolling resistance coefficient f	0.0494	/
Contacting track width L	3.57	m
Inertia of generator J_{gen}	2.0	kg·m^2
Inertia of engine J_{en}	3.2	kg·m^2
Gear ratio factor i_{eg}	1.6	/

of powertrain, road, and environment as

$$F_{ine} = \delta M_v a, \tag{2.2}$$

$$F_{rol} = f M_v g \cos \alpha, \tag{2.3}$$

$$F_{aer} = \frac{1}{2} C_d A \rho_{air} v^2, \tag{2.4}$$

$$F_{gra} = M_v g \sin \alpha, \tag{2.5}$$

where M_v is the vehicle curb weight and δ is the corresponding mass coefficient. f is the rolling resistance coefficient, a the vehicle acceleration, g the gravity acceleration and v the vehicle velocity. For the environment factors, A is the vehicle frontal area, C_d the aerodynamic drag coefficient, α the road slope angle and ρ_{air} the air density (the value is 1.25 kg/m^3 in normal conditions).

To compute the power demand of powertrain, the left and right sides of (2.1) times vehicle speed v is as follows:

$$P_{req} = F_{tra} v = (F_{ine} + F_{rol} + F_{aer} + F_{gra}) v. \tag{2.6}$$

Driving cycle represents both the way the vehicle is driven during a trip and road characteristics [67]. For simple expression, it can be understood as a time history of vehicle speed (and therefore acceleration) and road slope. Therefore, the power demand depends heavily on the driving cycles. Addressing an energy management problem can be addressed either by finding the optimal power split distribution for a special driving cycle or obtaining an adaptive and mutable policy for different driving cycles.

Furthermore, the primary difference between the wheeled vehicle and tracked vehicle is steering behavior. In the tracked vehicle, the rotative speed difference of the front wheels leads to the steering. However, the steering axle in the wheeled vehicle realizes this motion. Hence, the power requirement in hybrid tracked vehicle is different from (2.6) as

$$P_{req} = (F_{ine} + F_{rol} + F_{aer} + F_{gra})v + M_{\mu}\omega, \tag{2.7}$$

where M_{μ} is the resisting moment forced on the track and ω is the angular speed of the wheel.

From Fig. 1.2, the power provided by generator and battery are combined in the motor to drive the wheels. The power demand at the wheels is satisfied by energy sources with considering the energy conversion efficiency.

$$P_{req} = (P_{gen}\eta_{gen} + P_{bat}) \cdot \eta_{em}^{\pm 1}, \tag{2.8}$$

where P_{gen} and P_{bat} are the power of generator and battery, respectively. η_{gen} and η_{em} indicate the efficiency of the generator and EM. The positive signal of η_{em} represents propelling the vehicle and the negative one means regenerative braking.

2.1.2 ENGINE AND GENERATOR MODELING

For the energy management problem, the static map method is usually applied to model the ICE, wherein the fuel consumption rate is the significant variable. It is determined using a map (look-up table) as a function of the engine speed and torque, both calculated via powertrain dynamics,

$$\dot{m}_f = f(T_{en}, \omega_{en}), \tag{2.9}$$

where $\dot{m}_f$ is the fuel consumption rate, T_{en} the ICE torque and ω_{en} the ICE speed. The function f is affected by the engine size and capacity and is assumed to be immutable for a particular ICE. In addition, the engine torque is typically chosen as the control action in energy management, and its speed is measured by the driving cycle information.

Figure 2.1 depicts an example of the fuel consumption and efficiency maps for an engine. They are often obtained and evaluated by bench tests. The optimal operation line indicates that the corresponding torque and speed could obtain maximum efficiency. It is regarded as the target operating points to design the heuristics or rules in the energy management problem.

In series hybrid powertrain, the engine and generator are connected via a gearbox to meet the power demand at the wheels or charge the battery. The rated powers of the ICE and generator

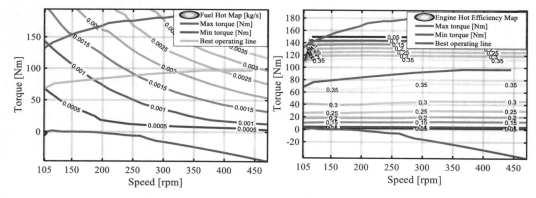

Figure 2.1: Overview of fuel consumption and efficiency maps in ICE modeling.

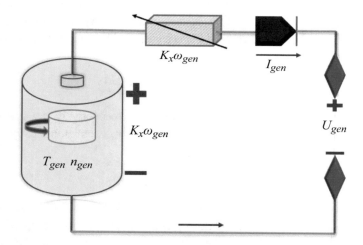

Figure 2.2: Generator modeling using an equivalent electric circuit approach.

are 300 kW and 270 kW, respectively. An equivalent electric circuit [68] is adopted to describe this connection, which comprises ICE, three-phase full-wave rectifier, and permanent magnet generator, as shown in Fig. 2.2. The engine has 2200 Nm rated output torque within the speed range [650, 2100] r/min and the generator has 960 Nm rated torque within the speed range [0, 2500] r/min [69].

The dynamics of voltage U_{gen} and current I_{gen} in the generator are expressed as

$$U_{gen} = K_e \omega_{gen} - K_x \omega_{gen} I_{gen} \tag{2.10}$$

$$K_e I_{gen} - K_x I_{gen}^2 = T_{gen}' \tag{2.11}$$

$$K_x = 3PL^g/\pi \tag{2.12}$$

$$P_{gen} = U_{gen}I_{gen}, \tag{2.13}$$

where $K_x\omega_{gen}$ is the electromotive force, K_e the electromotive force coefficient, and T_{gen} and ω_{gen} are the generator torque and speed. L^g is the synchronous inductance of the armature and P is the number of poles.

The torque equilibrium between the engine and generator is

$$\begin{cases} \frac{2\pi}{60}\left(\frac{J_{en}}{i_{ge}^2} + J_{gen}\right)\frac{d\omega_{gen}}{dt} = \frac{T_{en}}{i_{ge}} - T_{gen} \\ \omega_{en} = \omega_{gen}/i_{ge}, \end{cases} \tag{2.14}$$

where J_{en} and J_{gen} are the inertia moment of ICE and generator, respectively, and i_{ge} is the fixed ratio to connect these two components.

2.1.3   BATTERY MODELING

Dynamics of electrochemical components (battery or capacitor) are critical in hybrid electric vehicles because they influence driving range and output power. Many variables can characterize battery operation, such as temperature, voltage, state of charge (SOC), and current. SOC represents the current amount of electric charge stored in the battery and ranges from 0–100%. The objective of battery modeling in energy management problem is estimating the SOC variation as

$$S\dot{O}C(t) = -I_{bat}(t)/Q_c, \tag{2.15}$$

where Q_c is the battery nominal capacity and I_{bat} the battery current. Many approaches have been proposed to establish the battery modeling [70]; the internal resistance model is a representative and simplified one. The output power P_{bat} and voltage U_{bat} are decided as follows

$$\begin{cases} P_{bat} = U_{bat}I_{bat} \\ U_{bat} = V_{oc} - I_{bat}r_0, \end{cases} \tag{2.16}$$

where r_0 and V_{oc} indicate the internal resistance and open circuit voltage, respectively. They are changeable with the SOC, as displayed in Fig. 2.3.

Combining (2.15) and (2.16), the SOC differential is written as

$$S\dot{O}C(t) = \frac{(V_{oc} - \sqrt{V_{oc}^2 - 4r_0 P_{bat}(t)})}{2Q_c r_0}. \tag{2.17}$$

Two classic energy management problems are classified related to the final SOC value, which are charge sustaining (CS) and charge depleting (CD). CS mode means the final SOC value is close to its initial value in one journey. As a result, the battery and EM are only used to improve energy efficiency and power demand is primarily relied on by the engine. CD mode indicates the battery and EM mainly propel the vehicle, and thus the final SOC could reach

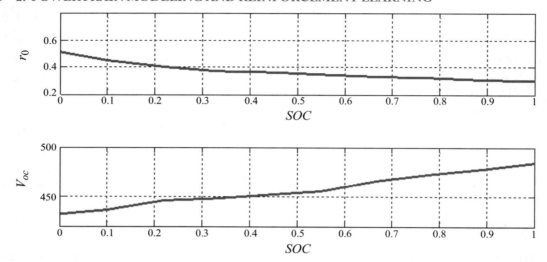

Figure 2.3: Variation of open circuit voltage and internal resistance with SOC.

a very low level. Most PHEVs operate in CD mode at startup and switch to CS mode (ICE launches) since the battery cannot support the power demand. Battery aging is represented as losing capacity and increasing internal resistance in a very long time horizon. Therefore, this book does not consider its dynamic effect in energy management.

2.1.4 EM MODELING

In series powertrain, the electric motor is directly connected to the final shaft and can work as motoring mode to propel the wheels or generating mode for regenerative braking. Its model can be built based on the efficiency related to the speed and torque. The mechanical output power P_{mech} is computed as follows

$$P_{mech} = T_{em} \cdot \omega_{em} = \begin{cases} \eta_{em} \cdot P_{elec} & motoring\,(P_{elec} \geq 0) \\ P_{elec}/\eta_{em} & generating\,(P_{elec} < 0)\,, \end{cases} \quad (2.18)$$

where T_{em}, ω_{em}, η_{em} are the torque, speed, and efficiency of EM, respectively. Obtaining the electric power P_{elec} generated by the EM is as follows:

$$P_{elec} = \begin{cases} (T_{em} \cdot \omega_{em})/\eta_{em} & motoring\,(P_{elec} \geq 0) \\ T_{em} \cdot \omega_{em} \cdot \eta_{em} & generating\,(P_{elec} < 0)\,. \end{cases} \quad (2.19)$$

The efficiency map of EM is also collected and verified by experiments. An example of this map is described in Fig. 2.4.

Efficiency Map

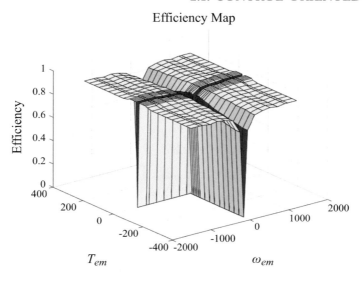

Figure 2.4: Efficiency map of EM in energy management problem.

2.1.5   ENERGY MANAGEMENT MODELING

Generally, the energy management problem in hybrid electric vehicles is always cast as an optimal control problem with a predefined objective and integral constraints. The goal is finding the best sequence of control actions to achieve the optimization of the cost function over a finite time horizon $[t_0, t_f]$. The cost function J (also called control objective or performance index) is usually formulated as

$$J = \varphi(x(t_f)) + \int_{t_0}^{t_f} L(x, a, t)dt, \tag{2.20}$$

where x and a are the state variable and control action and are time variant. L is the instantaneous cost function, which is a function of the state variable, control action, and time. In real applications, L could represent the exhaust emissions, fuel consumption, drivability, and thermal dynamics [71]. φ is a restrictive function on the final value of the state variable. In general, SOC is chosen as the state variable, and thus the above mentioned CD or CS problem can be displayed as

$$\varphi(x(t_f)) : x(t_f) - x(t_0) \approx 0 \tag{2.21}$$

$$\varphi(x(t_f)) : x(t_f) \in [SOC_{\min}, SOC_{\max}], \tag{2.22}$$

where $x(t_0)$ is the initial SOC value, and SOCmin and SOCmax are the minimum and maximum value of SOC (see Table 2.1). Equation (2.21) could represent the CS problem and (2.22) is the CD problem.

To calculate the instantaneous cost function at each time instant, the state variable should be acquired by using the control action, which results in the state equation as

$$\dot{x}(t) = f(x(t), a(t)). \tag{2.23}$$

Especially, as SOC is selected as the state variable, (2.17) is the state equation. Besides the constraint imposed on the final state variable, some local restraints are also necessary to guarantee the physical operation limits

$$T_{k,\min} \leq T_k(t) \leq T_{k,\max}, \ k = en, em, gen \tag{2.24}$$

$$\omega_{k,\min} \leq \omega_k(t) \leq \omega_{k,\max}, \ k = en, em, gen \tag{2.25}$$

$$P_{bat,\min} \leq P_{bat}(t) \leq P_{bat,\max} \tag{2.26}$$

where Equations (2.24) and (2.25) are the limitations for the engine, electric motor and generator speed, and torque. The engine torque or throttle is usually chosen as the control action in the energy management problem, which means the state variable and cost function are obtainable if they are known. If gear shifting is optimized in cost function or the powertrain is multimode, it is a good idea to limit the frequency of gear shifting or mode switching. Finally, the power demand at wheels should be satisfied by the onboard ESSs at each instant.

2.2   REINFORCEMENT LEARNING

Machine learning is like a subset of artificial intelligence; it enables computer systems to perform as intelligent machines by using statistical models or scientific algorithms. Three learning tasks are included in machine learning, which are unsupervised learning, supervised learning, and reinforcement learning [72]. Supervised learning indicates learning from a training set of labeled examples provided by a knowledgeable external supervisor. The trained system can extrapolate or generalize its responses based on the supervised examples. Different from supervised learning, there are no labeled outputs related to a set of inputs in an unsupervised learning task. The objective of this task is discovering patterns from the input data, such as grouping or clustering data points.

Reinforcement learning (RL) describes how to map situations to actions from the interaction between the environment and the agent. The learning process is a trial-and-error search, in which the agent should discover the best actions by trying them. Each action influences not only the immediate but also the subsequent rewards. These two features are the most representative characteristics of RL. Among all the formulations of machine learning, RL is the closest to imitating and learning what humans and other animals do, and many core algorithms of RL are inspired by biological learning systems. The formulation of an energy management strategy can be treated as the interaction procedure in RL. Regarding the mentioned challenges in energy management, this book aims to construct an RL-enabled intelligent energy management system for HEV. Due to its high computational efficiency and model-free characteristic, the relevant energy management policy is promising for use in real time.

2.2.1   OVERVIEW OF REINFORCEMENT LEARNING

As shown in Fig. 2.5, RL explicitly depicts the search problem as a goal-directed agent interacting with an uncertain environment [73]. Different RL agents have their own goals, which can sense aspects from their environments, and they can select actions to affect the environments. The agent needs to address the problem of how environment models are obtained and improved, as well as the question of how to maximize the cumulative rewards (the agent's objective) with the known knowledge of environments. Depending on if the knowledge of the environment is known or not, the search problems of RL are called planning and learning.

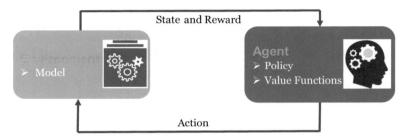

Figure 2.5: Interaction between agent and environment in RL.

Deriving or improving a policy based on a known environment model is called a planning or model-based RL; conversely, the trial-and-error learning problem needs to find the policy depending on the acquired experiences and is interpreted as model-free RL. Model-free RL algorithms focus on the fact that the agents need to spend more time on collecting experiences. Model-based ones can reserve the experiences first and exploit them to find the optimal policy. Hence, the model-free method requires less computational resources and the model-based one has higher calculative accuracy [74].

Besides the agent and environment, there are four more subelements in the RL framework. They are reward signal, value function, policy, and environment model. Reward signal represents the search objective of RL problem. At each time step, the agent chooses one action and the environment evaluates the performance of this selection by giving scalar signal called the reward as feedback. Hence, the reward signal defines what the good and bad events are for the agent. The target of the agent is to maximize or minimize cumulative rewards over a long time horizon. In general, the RL reward is usually a stochastic function of the environment's state and the agent's action.

As the RL agent cares about the accumulated rewards in the long run, value function in RL describes the total amount of reward an agent can receive from the current time step to the future. Therefore, the reward is the basics of the value function and the purpose of estimating values is to achieve more reward. That action may yield a low immediate reward but still generate high value. Thus, action choices should be made depending on the estimation of value. Actually, the core and emphasis of RL algorithms are estimating values of state accurately and efficiently.

A policy in RL is how a set of actions behaves with a given time sequence. It can seem to be the connection between the environment's state and the agent's action. Given a special state, the policy would give feedback on a particular action or stochastic one with probability. The goal of the RL agent is to search for an optimal policy to maximize the cumulative rewards. Note that the policy of RL may be sometimes a simple function or lookup table, whereas in others it may involve extensive computation.

Environment model mimics the reaction of the environment according to different actions. Generally, the environment model would generate the next state and reward via giving the current state and action. The relevant models are named transition model and reward model. As these models are known in advance, the planning technology is possible for policy deduction, which belongs to model-based methods. Model-free approaches depend on trial-and-error to gain experiences. In the next section, the materialization of RL elements is determined and analyzed in the energy management problem of HEV.

2.2.2   MARKOV DECISION PROCESSES

Markov decision processes (MDPs) are a typical formalization of sequential decision making, wherein actions affect not just immediate reward, but also subsequent rewards and states. MDP is a mathematically idealized form of the RL problem for which precise theoretical statements can be made, including reward, value function, and Bellman equations [75]. Regarding an RL problem, MDP is expressed as a quintuple $< S, A, P, R, \beta >$, wherein $s \in S$ and $a \in A$ are the state variable and control action, respectively, $p \in P$ and $r \in R$ are the transition model for state s and reward model for state-action pair (s, a), and β is a discount factor (the value of β is 0.95 in this book) to balance the importance of immediate and future rewards.

In the energy management problem of the series hybrid powertrain, SOC and generator speed are chosen as the state variables, and thus the state set is described as

$$S = \left\{ \omega_{gen}, SOC \mid 1200 \leq \omega_{gen}(t) \leq 3100, 0.2 \leq SOC(t) \leq 0.9 \right\}. \tag{2.27}$$

The engine torque is determined as the control action to compute the next state and reward through the powertrain dynamics:

$$A = \left\{ T_{en} \mid 0 \leq T_{en}(t) \leq 92 \right\}. \tag{2.28}$$

As discussed above, the reward signal represents the control objective in RL, which is included in the cost function in the energy management problem. Considering charge-sustaining mode as an example, the reward function is depicted as

$$\begin{cases} R = \left\{ r(s, a) : \dot{m}_f(s, a) + \theta \left(\Delta_{SOC} \right)^2 \right\} \\ \Delta_{SOC} = \begin{cases} SOC(t) - SOC_{ref} & SOC(t) < SOC_{ref}, \\ 0 & SOC(t) \geq SOC_{ref}, \end{cases} \end{cases} \tag{2.29}$$

where θ is a positive penalty factor to restrain the SOC value; SOC_{ref} is pre-allocated constant (usually equal to initial SOC value) to represent charge-sustaining constraint.

Finally, the transition model P is calculated as the transition probability matrix (TPM) of vehicle speed or power demand. For a particular driving cycle, the power demand is decided by Equation (2.6) or (2.7). A stationary Markov chain (MC) model is used to mimic the power demand, and the maximum likelihood estimator is applied to construct the TPM of power demand as follows:

$$\begin{cases} p_{ik,j} = P\left(P_{req} = P_{req}^j \,\middle|\, P_{req} = P_{req}^i, v = v_k\right) = \dfrac{N_{ik,j}}{N_{ik}} \\ N_{ik} = \displaystyle\sum_{j=1}^{M} N_{ik,j} \quad i, j = 1, 2, \ldots, M, \end{cases} \tag{2.30}$$

where M is the total amount of discrete index of power demand and k the discrete time step. N_{ik} is the total transition counts initiated from P_{req}^i at vehicle speed v_k, and $N_{ik,j}$ indicates the number of transitions from P_{req}^i to P_{req}^j that happened at vehicle speed v_k. Figure 2.6 displays an example of TPM of power demand in the energy management problem of HEV.

Since TPM could represent the transition dynamics of vehicle speed or power demand, the differences between two TPMs can be employed to identify the corresponding differences in speed or power demand. Two technologies are proposed to quantify these differences, which are Kullback–Leibler (KL) divergence rate and induced matrix norm (IMN).

Assuming P_1 and P_2 are two TPMs of power demand, the KL divergence rate is described as [76]

$$D_{KL}\left(P_1 \,\|\, P_2\right) = \sum_x \sum_{x^+} [P_1(x^+|x)P^*(x)] \log\left[\frac{P_1(x^+|x)}{P_2(x^+|x)}\right], \tag{2.31}$$

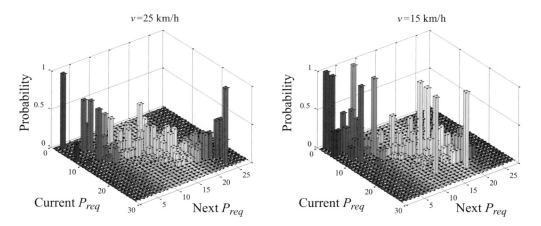

Figure 2.6: Instance of TPM of power demand in energy management.

where x and x^+ are the indices of transition probability, and they range from 1 to M. P^* is the steady-state probability distribution of P, which is determined by

$$P^* P_1 = P^*. \tag{2.32}$$

It is obvious that P^* is an eigenvector of P_1 whose eigenvalue corresponds to 1.

Furthermore, the IMN is also introduced to measure the similarity of two matrices

$$IMN\left(P_1 \| P_2\right) = \| P_1 - P_2 \|_2 = \sup_{x \in R^M / \{0\}} \frac{|(P_1 - P_2)x|}{|x|}, \tag{2.33}$$

where x depicts an $M \times 1$ dimension non-zero vector, and sup is the supremum of a scalar. To simplify computation, (2.33) can be reformulated as

$$IMN\left(P_1 \| P_2\right) = \| P_1 - P_2 \|_2 = \max_{1 \le i \le M} |\lambda_i (P_1 - P_2)|$$

$$= \max_{1 \le i \le M} \sqrt{\lambda_i ((P_1 - P_2)^T (P_1 - P_2))}, \tag{2.34}$$

where $\lambda_i(P)$ is an eigenvalue of a matrix for arbitrary $i \in [1, M]$, and P^T denotes the transpose of a matrix P. Note that for KL divergence rate or IMN, the closer the value is to zero, the more similar the TPM P_1 is to P_2.

2.2.3   ALGORITHMS FOR RL: Q-LEARNING AND SARSA

In the following two sections, four RL algorithms are formulated and explained to derive the energy management policy for the series hybrid powertrain, which are Q-learning, Sarsa, Dyna-Q, and Dyna-H. After the specialization of the MDP in RL, two kinds of value functions are generally established by the related elements as

$$V^\pi (s) = E \left\{ \sum_{t_0}^{t_f} \beta r | s \right\} \tag{2.35}$$

$$Q^\pi (s, a) = E \left\{ \sum_{t_0}^{t_f} \beta r | s, a \right\}, \tag{2.36}$$

where π is the control policy in RL. Equation (2.35) is called a state value function and (2.36) is the action-value function. The difference between these two functions is whether the current action is known or not. Thus, the action value function is often used to derive the optimal action at each time step. To realize this, (2.36) could be rewritten as the Bellman form

$$Q^\pi (s, a) = r(s, a) + \beta \sum_{s' \in S} p_{s'a,s} Q^\pi (s', a'), \tag{2.37}$$

where $p_{s'a,s}$ denotes the transition probability from s to s' by taking action a. Then, an optimal control action is decided by maximizing or minimizing the value function

$$\pi^*(s) = \arg\max_a \left[r(s,a) + \beta \sum_{s' \in S} p_{s'a,s} Q^\pi(s',a') \right]. \tag{2.38}$$

For the two value functions, the following expression is satisfied, $V^*(s) = Q^*(s,a)$. It is obvious that the optimal control policy is completely determined by the action value matrix $Q(s,a)$. Moreover, this matrix is decided by the control action at each time step. The ε-greedy policy is usually used to select control actions. It represents that the agent explores a random control action with probability ε to increase the experiences, and exploits the best action in $Q(s,a)$ matrix until now with probability $1-\varepsilon$.

In RL algorithms, if the current and next actions are all chosen by ε-greedy policy, this algorithm is called on-policy. Conversely, if the next action is not selected by ε-greedy policy, it is named off-policy. The typical off-policy and on-policy algorithms are Q-learning and Sarsa. Their updating equation are depicted as follows [77, 78]:

$$Q(s,a) \leftarrow Q(s,a) + \alpha \left[r + \beta \max_{a'} Q(s',a') - Q(s,a) \right] \tag{2.39}$$

$$Q(s,a) \leftarrow Q(s,a) + \alpha [r + \beta Q(s',a') - Q(s,a)], \tag{2.40}$$

where $\alpha \in [0,1]$ is a learning rate to trade off the old and newly learned knowledge. The next action in (2.39) is selected via the maximum probability and is chosen by the ε-greedy policy in (2.40). ε-greedy policy indicates spending more time to collect experiences and find the environment model. Maximum probability means selecting the next action depends on the current best knowledge. Hence, the Sarsa may consume more time; however, its performance may be better than Q-learning.

To describe the realization process of the RL algorithm, Table 2.2 shows the pseudo-code of the Q-learning algorithm, which is easily changed as Sarsa algorithm by replacing step 6. The calculative episodes are 100 and, in each episode, the iteration number K is 1000. The learning rate α is changed with time instant and decided as $1/\sqrt{k+2}$, and the sample time is 1 s.

2.2.4   ALGORITHMS FOR RL: DYNA-Q AND DYNA-H

In addition to the common RL algorithms, Q-learning and Sarsa, this section formulates Dyna-Q and Dyna-H to promote the learning process more accurately and efficiently. Dyna-H synthesizes three RL processes, which are model learning, heuristic planning, and Q-learning, as depicted in Fig. 2.7. Model learning signifies recording the models (s,a,s',r) after undergoing experiences. Planning indicates producing random state-action pair (s,a) from the stored experiences, which can improve the learning efficiency of control policy [79]. However, choosing experiences randomly may not be the best way; as an alternative, focusing on special state-action pairs could achieve more efficient searching [80].

Table 2.2: Pseudo-code of Q-learning algorithm

Q-learning Algorithm
1. Initialize action value matrix $Q(s, a)$, state s, and number of iteration K
2. Repeat each time step k ($k = 1, 2, 3...K$)
3. Determine a, depend on $Q(s, \cdot)$ (ε-greedy policy)
4. Taking action a and observe reward r, next state s'
5. Search optimal control action $a^* = \arg\max_a Q(s', a)$
6. $Q(s, a) \leftarrow Q(s, a) + \alpha(r(s, a) + \beta\max_{a'} Q(s', a') - Q(s, a))$
7. $s \leftarrow s'$
8. until s is terminal

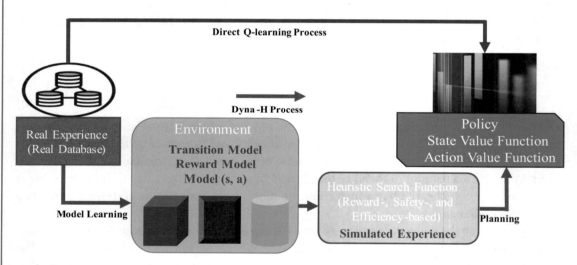

Figure 2.7: Integrated learning and heuristic planning architecture: Dyna-H.

Dyna-Q algorithm is a typical RL architecture in which the online planning and model learning are integrated together. Here, in this book, direct learning represents the Q-learning algorithm and indirect learning indicates Dyna-Q, which includes (planning process inside). Model learning usually means deciding the next state and reward through table-based methods. During planning, the algorithm randomly samples only state-action pairs that have been experienced previously [81].

The relevant pseudo-code of Dyna-Q algorithm is described in Table 2.3. *Model*(s, a) represents the predicted next state and reward by state equations for the state-action pair (s, a). N is the total planning step in each episode. Note that steps 6, 7, and 8 represent the above mentioned direct learning, model learning and planning processes. As long as steps 7 and 8 are neglected, the remaining algorithm expresses the Q-learning algorithm.

Table 2.3: Realization process of Dyna-Q algorithm

Dyna-Q Algorithm
1. Initialize $Q(s, a)$, $Model(s, a)$, for each state $s \in S$, action $a \in A$
2. repeat {in each episode}
3. $s \leftarrow$ current (not terminal) state
4. $a \leftarrow \varepsilon\text{-}greedy$ decides actions (s, Q)
5. implement a; observe next state s' and reward r
6. $Q(s, a) \leftarrow Q(s, a) + \alpha[r(s, a) + \beta\max_{a'} Q(s', a') - Q(s, a)]$
7. Model learning $Model(s, a) \leftarrow s', r$
8. for $i = 1$ to N do (Planning process)
9. $s \leftarrow$ random choose previously observed state
10.   a $\leftarrow$ random choose action previously taken in s
11. $s', r \leftarrow$ Model learning $Model(s, a)$
12. $Q(s, a) \leftarrow Q(s, a) + \alpha[r(s, a) + \beta\max_{a'} Q(s', a') - Q(s, a)]$
13.   end for
14. until s' is terminal

To speed up the search process of power split controls for the series powertrain, the Dyna-H algorithm is formulated by incorporating the Dyna-Q with a heuristic planning strategy. This heuristic planning strategy is called heuristic function H. In this book, the Euclidean distance measurement is used to represent the heuristic function, wherein the inputs are the current reward and a defined goal reward [82]

$$H(s, a) = \left\| r(s, a) - r_{goal} \right\|^2, \tag{2.41}$$

where r_{goal} is the goal reward. The calculative workflow of Dyna-H algorithm is expressed in Table 2.4. Steps 6 and 7 are Q-learning and model learning, and steps 8–14 represent the heuristic planning process. $Model(s, a)$ represents the learned experiences based on the state-action pair (s, a) [82]. The worst action would be chosen for each state at step 9 and ignored in the following episode to improve search speed, wherein $H(s, a)$ denotes the heuristic chosen rule of control action. Finally, the mentioned RL algorithms are implemented in Matlab using a 2.90 GHz microprocessor with 7.83 GB RAM. The relevant RL toolboxes are MDP toolbox [83] and Dyna-H toolbox [81].

Table 2.4: Realization process of Dyna-H algorithm

Dyna-H Algorithm
1. Initialize $Q(s, a)$, *Model*(s, a), N, for each state $s \in S$, action $a \in A$
2. repeat{for each episode}
3. $s \leftarrow$ current (not terminal) state
4. $a \leftarrow$ *ε-greedy* decides actions (s, Q)
5. execute a; observe next state s' and reward r
6. $Q(s, a) \leftarrow Q(s, a) + \alpha[r + \beta\max_{a'} Q(s', a') - Q(s, a)]$
7. Model learning *Model*$(s, a) \leftarrow s', r$
8. for $i = 1$ to N do (Planning process)
9. $a \leftarrow H(s, a)$
10.   if $s, a \notin$ *Model* then
11. $s \leftarrow$ random choose previously observed state
12. $a \leftarrow$ random choose action previously taken in s
13.   end if
14. $s', r \leftarrow$ Model learning *Model*(s, a)
15. $Q(s, a) \leftarrow Q(s, a) + \alpha[r + \beta\max_{a'} Q(s', a') - Q(s, a)]$
16. $s \leftarrow s'$
17. end for
18. until s' is terminal

2.3 SUMMARY

The performance and effectiveness of energy management are highly reliant on the accuracy of powertrain modeling in HEV. As a matter of fact, the target series hybrid electric powertrain is built to illuminate the power flow and the dependency relationship of different ESSs. To solve this energy management problem, it is naturally transferred into an optimization control problem including state variables, control actions, optimization objective, and physical constraints. Taking a series powertrain as an example, the RL architecture is formulated to handle this optimization control problem due to its high calculative efficiency and model-free characteristic. In an effort to construct an intelligent energy management system, four RL algorithms are explained and their related realization processes are labeled. By integrating the online and predictive driving cycle and power demand information (introduced in the next chapter), the mentioned RL algorithms are promising ways to derive adaptive and efficient EMSs. The effectiveness and correctness of the intelligent energy management system is also evaluated and validated in the following chapters.

CHAPTER 3

Prediction and Updating of Driving Information

3.1 PREDICTIVE ALGORITHMS

Future driving information (vehicle speed or power demand) is extremely significant for online and adaptive energy management system formulation. When future vehicle velocity and power demand are able to be predicted accurately, the corresponding EMS could be regulated appropriately a priori to accommodate driving conditions. Especially when driving situations are switched (e.g., from the highway to urban), the predicted driving information is important for an intelligent learning-based energy management system. With this goal in mind, many algorithms have been proposed to forecast driving cycles in HEV's energy management problem, such as k-nearest neighbor (k-NN), artificial neural network (ANN), Markov chain (MC), support vector machine (SVM), etc. For example, the authors in [84, 85] used MC to simulate the driving cycles and then combined them with particle filter or fuzzy logic rules to derive the predictive EMSs. The experiment tests indicate the short-term driving cycles prediction could effectively improve control performance in different cost functions. According to this historical data, the future driving cycle information could easily be obtained from database search [86]. For large-scale driving data, SVM is regarded as an efficient tool to recognize and classify predetermined features, and then forecast the to-be vehicle velocity and road slope [87].

Along with deep learning widely used in many research areas, ANN is considered a promising solution to achieve precise prediction for a time sequence. For the energy management problem, the authors of Refs. [88, 89] tried to apply different types of the neural network to realize vehicle velocity prediction. For example, Feng et al. [88] and Xiang et al. [89] selected radial basis function neural network to get the future power demand and vehicle speed, respectively. These novel predictive EMSs reflect better performance via comparing with ECMS and PID controls. Furthermore, Refs. [90] and [91] employed another two algorithms for short-term driving cycles prediction, which are the car-following model and MPC. Based on the particular models, future speed information can be easily obtained to promote performance on many control objectives. By predicting future driving cycles, the energy management controller is able to not only save fuel consumption but also enhance the safety and stability of the powertrain.

In this section, the MC models and deep learning method are utilized to achieve vehicle speed or power demand prediction. For MC models, two specific approaches, nearest neighborhood (NND) and fuzzy coding (FCG), are studied to realize driving cycle prediction. For deep learning direction, the popular and powerful technique, long short-term memory (LSTM), is applied for vehicle speed prediction. The constructed predicted framework can be easily transformed to achieve power demand prediction when the historical data is available. Actually, the proposed three methods are suitable for arbitrary time sequence prediction of the hybrid powertrain, such as engine speed, generator torque battery power, etc. The obtained driving information is regarded as the inputs for RL-based intelligent energy management system, which would result in online and adaptive EMSs for various driving conditions.

3.1.1   NEAREST NEIGHBORHOOD

To predict the time sequence (e.g., driving information) in hybrid electric vehicles, the variable interval is divided into finite discrete states, which are represented as $X = \{x_i | i = 1, \ldots, M\}$ and subject to $\{x_i < x_{i+1} | \forall i\}$. For an arbitrary continuous time sequence, it can be mapped to these discrete states via nearest neighborhood measurement as follows:

$$y = x_j, \quad j = \arg\min_i |y - x_i|, \tag{3.1}$$

where y denotes a special time sequence at arbitrary time point. After the mapping is established, a Markov chain (MC) is usually used to mimic the discrete states. Then, the transition probability of these states is computed via maximum likelihood estimator (MLE) as

$$p_{ij} = \frac{N_{ij}}{N_{io}}, \quad i, j \in \{1, \ldots, M\} \tag{3.2}$$

where p_{ij} represents the transition probability from x_i to x_j. N_{ij} and N_{io} are the transition counts from x_i to x_j and x_i to all the objectives, respectively. Thus, these two numbers are satisfied by

$$N_{io} = \sum_{j=1}^{M} N_{ij}, \quad i, j \in \{1, \ldots, M\}. \tag{3.3}$$

The obtained example of transition probability is shown in Fig. 3.1. Based on the obtained transition probability, the future one-step and multi-step variables can be predicted in two forms. The first one is through the maximum probability to compute the future variable as

$$x_{t+1} = x_k, \quad \text{if} \quad x_t = x_i, \quad k = \arg\max_k (p_{ik}). \tag{3.4}$$

The second idea is exploiting the expectation of the next state to depict the predicted value as [92]

$$x_{t+1} = \sum_{k=1}^{M} p_{ik} \cdot x_k, \quad \text{if} \quad x_t = x_i. \tag{3.5}$$

It is obvious that the second expression is more accurate than the first one because it considers all possible transitions. In this book, the second formulation is utilized to represent

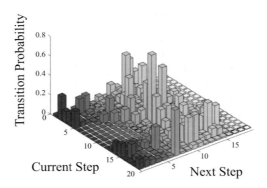

Figure 3.1: Example of computed transition probability for a time sequence.

the NND-based prediction. Finally, the multi-step future variables are computed by extending (3.5) as follows:

$$x_{t+n} = \sum_{k=1}^{M} (p_{ik})^n \cdot x_k, \quad \text{if} \quad x_t = x_i. \tag{3.6}$$

3.1.2   FUZZY CODING

In FCG method, the discrete signal states x_i are replaced by fuzzy sets A_i, $i = 1, \ldots, M$. Each set A_i is a pair of X and μ_i, in which μ_i means the Lebesgue measurable membership function. Its definition is shown below:

$$\mu_i : X \to [0,\ 1], \quad s.t. \quad \forall y, \ \exists i, \ \mu_i(y) > 0. \tag{3.7}$$

Hence, $\mu_j(y)$ suggests the degree of membership of $y \in X$ in μ_j. Unlike interval partitioning in NND, a continuous state $y \in X$ in the fuzzy encoding method may be associated with several states μ_i with different degrees of membership.

Since this membership maps X to $[0,\ 1]$, a transformation could be executed from arbitrary variable y to a possibility vector $\mu^T(y)$. The transformation allocates each $y \in X$ with an M-dimensional possibility (not probability) vector as follows:

$$\mu^T(y) = [\mu_1(y), \mu_2(y), \ldots, \mu_M(y)]. \tag{3.8}$$

As the $\mu_i(y)$ represents the degree of membership, the sum of the vector $\mu^T(y)$ may not equal 1 because each variable y could be associated with more than one fuzzy set A_i. This transformation is named as fuzzification, which aims to map each time sequence point of the space X to an M-dimensional possibility vector in the space $\tilde{X}$.

Then, another transformation called the proportional possibility-to-probability transformation is used to convert the possibility vector $\mu^T(y)$ to a probability vector $\xi(y)$ by normalization [93]

$$\xi_i(y) = \mu_i(y) / \sum_{j=1}^{M} \mu_j(y), \quad i = 1, \dots M, \tag{3.9}$$

where this transformation connects the possibility space $\tilde{X}$ with an M-dimensional probability vector space, $\bar{X}$. Since the TPM P of time sequence X is calculated by Equations (3.2) and (3.3), the next state in space $\bar{X}$ is described as follows:

$$(\xi_i(y))' = \xi_i(y) \cdot P \tag{3.10}$$

where x' indicates the next state of x. In (3.10), the element p_{ij} in the TPM P can be interpreted as the transition of fuzzy sets, from A_i to A_j.

Finally, to realize the goal of predicting the next state in space X, a decoding process is necessary via aggregating the membership function $\mu^T(y)$ with probability distribution $\xi(y)$ as [94]:

$$q'(y) = (\xi(y))' \mu(y) = \xi(y) \cdot P\mu(y). \tag{3.11}$$

For the continuous space, the next state x' is also computed as the expected value over the possibility vector

$$\begin{cases} x' = \int_X q'(y) y \, dy / \int_X q'(y) \, dy \\ \int_X q'(y) y \, dy = \sum_{i=1}^{M} \mu_i(y) \sum_{j=1}^{M} p_{ij} \int_X y \mu_j(y) \, dy \\ \int_X q'(y) \, dy = \sum_{i=1}^{M} \mu_i(y) \sum_{j=1}^{M} p_{ij} \int_X \mu_j(y) \, dy. \end{cases} \tag{3.12}$$

Define the centroid and volume of the membership function $\mu_j(y)$ as follows:

$$\begin{cases} \bar{c}_i = \int_X y \mu_j(y) \, dy \\ V_j = \int_X \mu_j(y) \, dy. \end{cases} \tag{3.13}$$

Note that the sum of the probability vector is equal to 1, which means $\sum_{j=1}^{M} p_{ij} = 1$ and $\sum_{i=1}^{M} \mu_i(y) = 1$. Equation (3.12) is able to be simplified as

$$\begin{cases} x' = \dfrac{\sum_{i=1}^{M} \mu_i(y) \sum_{j=1}^{M} p_{ij} \bar{c}_j}{\sum_{i=1}^{M} \mu_i(y) \sum_{j=1}^{M} p_{ij}} = \mu(y)^T P \bar{c}. \end{cases} \tag{3.14}$$

Obviously, the next state in FCG technique is strongly related to the membership functions. In this book, the Gaussian membership function with standard deviation $\sigma = 1$ is considered to forecast the practical vehicle speed as

$$\mu_i = e^{\frac{-(x-2.5i+1.25)^2}{2\cdot\sigma^2}}, \quad i = 1,\ldots,M. \tag{3.15}$$

Overall, both the NND and FCG methods depend on the MC models, which implies that the TPM is necessary for these two approaches. For different collected driving data, the TPMs are essentially different, which leads to the various prediction results. Thus, the accuracy of these two MC-based prediction methods relies on the precision of the collected historical data. The performance of these two methods is compared and analyzed in the following section.

3.1.3   LONG SHORT-TERM MEMORY

Deep learning (DL) has been a hot research topic in recent years, inspired by learning features from collected data representations. DL usually contains several levels, and each level aims to transfer the coarse data into a composite and abstract representation. Generally, the artificial neural network (ANN) is used to mimic the DL models and can be indicated as function approximation between the input and output. This approximation is often achieved through three common layers, which are input, hidden, and output layers [95].

The neuron is a sole computation unit of ANN; see Fig. 3.2 as an illustration. The input and weight vectors are denoted as $a = [a_1, a_2, \ldots, a_N]$ and $\omega = [\omega_1, \omega_2, \ldots, \omega_N]$, and then the neuron input b is depicted as

$$b = \sum_{i=1}^{N} a_i \cdot \omega_i + B \tag{3.16}$$

where B is a bias. The activation function is acted on the neuron input to generate the restricted and squashed output value

$$c = f(b), \tag{3.17}$$

wherein the activation function f can be Sigmoid, Tanh and Rectified Linear Unit (Relu) function.

In many cases, the order of inputs will influence the outputs, such as the natural language data, speech, and music sequence data. Recurrent neural network (RNN) is conducted to address this problem, in which the hidden information will be delivered to the next step as time passes, as shown in Fig. 3.3. U and V are the weight vectors for input and output, and W is the weight for each time step. The calculative process of RNN at time step t is described as follows:

$$H_t = f_1 (a_t \cdot U_t + H_{t-1} \cdot W) \tag{3.18}$$

$$c_t = f_2 (H_t \cdot V_t) \tag{3.19}$$

where H_{t-1} and c_{t-1} are decided at time step $t - 1$. Note that all the weights in RNN are initialized first, and then they are adjusted through back propagation to decrease the error.

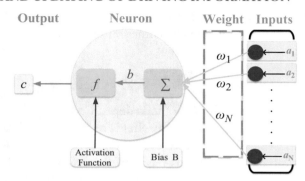

Figure 3.2: The architecture of the neuron unit in ANN.

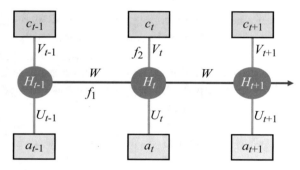

Figure 3.3: The workflow diagram of RNN.

In the backpropagation of RNN, the gradient descent algorithm is usually utilized to update the values of weight. However, the vanishing or exploding gradient problem happens frequently to stop RNN from being used in long-range time sequence. As an alternative, long short term memory (LSTM) contains a memory cell to store the extra information.

The diagram of the LSTM network is depicted in Fig. 3.4. Six key components are included in this network, which are forget gate F, candidate layer G, input gate I, output gate O, hidden state H, and memory state M. The inputs of this network at time step t are the past hidden and memory states, and the current input X_t. The outputs are the fresh memory and hidden states. By transiting the memory state, the LSTM network is able to remember arbitrary time intervals' series. The signs of σ and $\tanh$ denote the sigmoid and Tanh activation function, respectively. W and U are the weights for three gates and candidate layer with special subscripts.

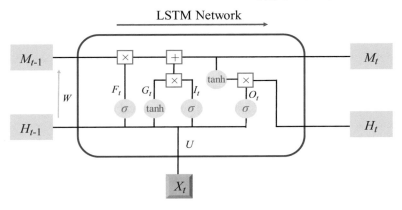

Figure 3.4: The construction of the LSTM network.

According to the RNN's computational process, the expressions of LSTM can be summarized as follows:

$$F_t = \sigma \left(X_t \cdot U_f + H_{t-1} \cdot W_f \right) \tag{3.20}$$

$$G_t = \tanh \left(X_t \cdot U_c + H_{t-1} \cdot W_c \right) \tag{3.21}$$

$$I_t = \sigma \left(X_t \cdot U_i + H_{t-1} \cdot W_i \right) \tag{3.22}$$

$$O_t = \sigma \left(X_t \cdot U_o + H_{t-1} \cdot W_o \right). \tag{3.23}$$

Finally, the outputs of the LSTM at time step t can be expressed as a function of the resulting variables

$$M_t = F_t \cdot M_{t-1} + I_t \cdot G_t \tag{3.24}$$

$$H_t = O_t \cdot \tanh \left(M_t \right). \tag{3.25}$$

After training the network with collected experienced experiment data, the learned model could provide new predictions according to new inputs. In this book, the LSTM for time sequence prediction is realized in Matlab 2018b using the deep learning toolbox.

Finally, since the discussed NND and FCG approaches are both dependent on the transition probability of the studied time sequence, they will be compared to certify the predicted performance. Furthermore, one-step and multi-step performances of the LSTM network are compared and discussed, and the differences between the observed and predicted trajectories are quantified by the root mean square error (RMSE)

$$RMSE = \sqrt{\sum_{i=1}^{M} \left(x_{pre}^i - x_{ob}^i \right)^2 / M} \tag{3.26}$$

where x_{pre} and x_{ob} are the forecast and observed time sequence, respectively.

3.2   ONLINE UPDATING ALGORITHM

Reviewing the parameters in the RL framework (state, action, reward, value function, and transition model) shows the transition model is a potential and mutable element. As the driving information (vehicle speed or power demand) is also changeable in the energy management problem of HEV, the transition model is a promising method to reflect the variation of driving information. Specifically, the TPM of the vehicle velocity or power demand can be changed with driving conditions to represent the state transition in RL. Hence, this section addresses how to compute (update) the TPM when the driving situations change.

In energy management, the TPMs would be updated in two perspectives. First, when the current driving condition is greatly different from historical driving data, the old TPM would be replaced by the current TPM. By doing this, the relevant RL-based power split control is altered because the transition model has been changed. The second situation is related to the predicted driving information. Assuming the future power demand or vehicle speed is predicted by the above three approaches, the comparison between current and future information is feasible. If the difference is huge, the energy management system should change its strategy in time to adapt to the future driving conditions. This renewal of energy management policy is able to be achieved by updating the TPM computation.

To update the TPM online, it is convenient to modify (3.2) into a real-time application form as

$$p_{ij} = \frac{N_{ij}(L)}{N_{oi}(L)} = \frac{N_{ij}(L)/L}{N_{oi}(L)/L} = \frac{F_{ij}(L)}{F_{oi}(L)} \quad i,j \in \{1,\ldots,M\} \tag{3.27}$$

where L represents the current or future time horizon of the considered time sequence. $F_{ij}(L)$ indicates the frequency rate of transition events $f_{ij}(L)$, which means transferring from x_i to x_j. $F_{oi}(L)$ is the total frequency rate to describe the transition events $f_{oi}(L)$ initiated from x_i. As the time length L changes, these two frequency rates would alter as well. The relationship between the frequency rate and transition events is depicted below:

$$\begin{cases} F_{ij}(L) & = \frac{1}{L} \displaystyle\sum_{t=1}^{L} f_{ij}(t) \\ F_{oi}(L) & = \frac{1}{L} \displaystyle\sum_{t=1}^{L} f_{oi}(t) = \frac{1}{L} \displaystyle\sum_{t=1}^{L} \sum_{j=1}^{M} f_{ij}(t) \end{cases} \tag{3.28}$$

where the value of $f_{ij}(t)$ and $f_{oi}(t)$ is equal to 1 if and only if a transition from x_i to x_j and a transition initiated from the state x_i occur at time instant t, respectively. Otherwise, they take values of zero. To update the transition probability, the two expressions in (3.28) can be written as iterative forms

$$F_{ij}(L) = \frac{1}{L} \sum_{t=1}^{L} f_{ij}(t) = \frac{1}{L} \left[(L-1) F_{ij}(L-1) + f_{ij}(L) \right]$$
$$= F_{ij}(L-1) + \frac{1}{L} \left[f_{ij}(L) - F_{ij}(L-1) \right]$$
$$= F_{ij}(L-1) + \gamma \left[f_{ij}(L) - F_{ij}(L-1) \right]$$

(3.29)

$$F_{oi}(L) = \frac{1}{L} \sum_{t=1}^{L} f_{oi}(t) = \frac{1}{L} \left[(L-1) F_{oi}(L-1) + f_{oi}(L) \right]$$
$$= F_{oi}(L-1) + \frac{1}{L} \left[f_{oi}(L) - F_{oi}(L-1) \right]$$
$$= F_{oi}(L-1) + \gamma \left[f_{oi}(L) - F_{oi}(L-1) \right]$$

(3.30)

where $\gamma = 1/L$. To weight the old time sequence data with exponentially decreasing weights for online application, this constant parameter is replaced by a value ranging from 0–1, which means $\gamma \in (0, 1)$.

It is named the forgetting factor to determine the effective memory depth and control updating rate of the TPM [94, 96].

Finally, the online updating formulation of TPM for arbitrary driving information in the energy management problem of HEV is determined as [56]

$$p_{ij} = \frac{F_{ij}(L)}{F_{oi}(L)} = \frac{F_{ij}(L-1) + \gamma \left[f_{ij}(L) - F_{ij}(L-1) \right]}{F_{oi}(L-1) + \gamma \left[f_{oi}(L) - F_{oi}(L-1) \right]}.$$

(3.31)

Note that the time horizon L suggests how often the controller considers updating the TPM. In this book, for the energy management of series powertrain, the value of L is 100 seconds by considering the computational efficiency and practical significance simultaneously.

3.3   EVALUATION OF PREDICTION PERFORMANCE

The prediction performance of NND, FCG, and LSTM is estimated in this section. Taking vehicle speed as an example, the evaluation is conducted in three aspects. First, the importance of the historical data in NND is illuminated, which indicates the NND method needs to rely on the empirical data to achieve accurate predictions. Then, a comparative analysis between NND and FCG is constructed, and their positives and negatives are described. Finally, the performance of the LSTM network is displayed by showing one-step and multi-step predicted results.

3.3.1   NND-ENABLED PREDICTION RESULTS

To apply the NND algorithm to forecast the speed trajectory, the transition probability is initialized as zero before the prediction. It will be mature as long as the vehicle goes through enough driving cycles. As the time step increases, the transition probability matrix becomes more mature. Figure 3.5 depicts three rounds of prediction for the same cycle, which means the NND method is used to forecast this cycle three times. It is obvious that the performance of the second round is close to that of the third round, and both are better than the first round. In some cases, the errors are very large in the first round and they decrease in the second and third round. This owes to the update of the transition probability matrix, which indicates the probability is not stable (still the initial values) in the first round. These results indicate that NND approach

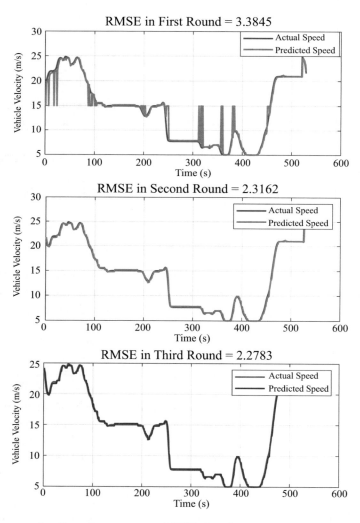

Figure 3.5: Three rounds of prediction using NND for the same cycle.

could achieve good performance based on adequate experiences. Therefore, it is suitable for driving information prediction of vehicles with fixed routes, such as bus, sanitation vehicle, and the garbage truck.

For multi-step prediction, Fig. 3.6 displays the NND-based 10-step prediction in the first and second rounds. The blue line is the observed value, and the red lines are the 10-step predicted values at each time instant. By comparing the RMSE values in Equation (3.26), it can be discerned that the performance of the second round is better than the first round. Hence, the NND technique needs similar experienced data to improve and complete the transition probability. After that, it can achieve good performance in one-step and multi-step predictions.

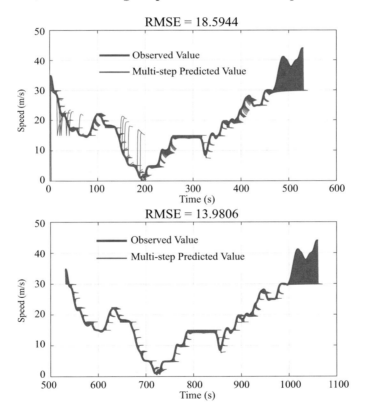

Figure 3.6: NND-based 10-step prediction in the first and second round.

3.3.2 COMPARISON OF NND AND FCG

Since the TPM is necessary in NND and FCG, we compare them to discern the differences between them. The one-step forecast speed trajectories of these two methods are depicted in Fig. 3.7. Based on the curves and RMSE, it can be seen that the performance of FCG is better

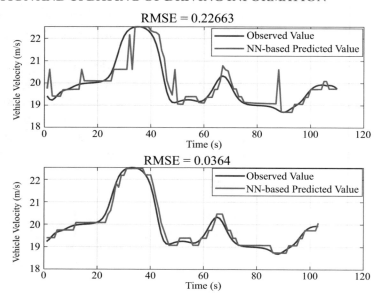

Figure 3.7: Compared speed trajectories between NND and FCG in the first round.

than that of NND, even in the first round. This is a result of the division of the state space X in these two methods. In NND, X is divided into serval discrete states, and thus speed transition at each time step only leads to one row updating in TPM. However, for FCG, X is segmented into many fuzzy sets and each time sequence point may be related to multiple fuzzy sets. Hence, each speed transition may cause many rows' updating or even the whole TPM's updating. From the mathematics point of view, FCG may cost more time than NND but would generate better prediction results.

To intuitively exhibit the differences between these two methods, Fig. 3.8 shows the TPMs in these methods at the same time step. For NND, some high-speed transitions have not been experienced, and thus the transition probabilities are still initial values. Conversely, the matrix of FCG becomes stable owing to the updating rule. Besides the better predicted performance, the FCG approach also has a drawback when compared with NND. The computation time of NND is about 2 seconds. However, it will cost 500 seconds to finish prediction in FCG. Hence, FCG is not suitable for online prediction. To forecast some variables for hybrid vehicles, the NND method is feasible to predict the known driving situations online where the trajectory has been experienced, and the FCG is appropriate to predict the unknown environments offline to achieve higher accuracy.

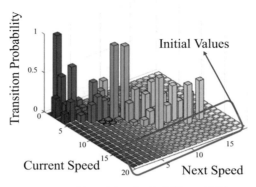

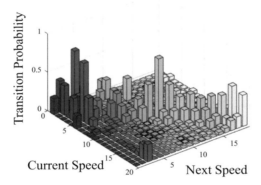

Figure 3.8: Computed transition probability in NND and FCG.

3.3.3   EVALUATION OF LSTM

LSTM network for prediction is realized by using the *predictAndUpdateState* function in the deep learning toolbox. In the operational process, the collected historical driving cycles are imported to train the network first, and then this learned model can generate new predictions by giving new inputs. The specialization of the parameters in LSTM is defined, where the training episode is 150, the number of hidden units is 100, the learning rate is 0.005, and the gradient threshold is 1. Figure 3.9 describes the training process of the LSTM network, which is generated by the toolbox and contains trajectories of RMSE and loss function. These two values drop along with the iteration number, which results in improved accuracy.

To explain the significance of the training data in LSTM network, Fig. 3.10 shows the predicted values based on different historical data. (a) and (c) indicate the multi-step prediction (about 40 steps prediction) with different training data, and (b) and (d) are the one-step

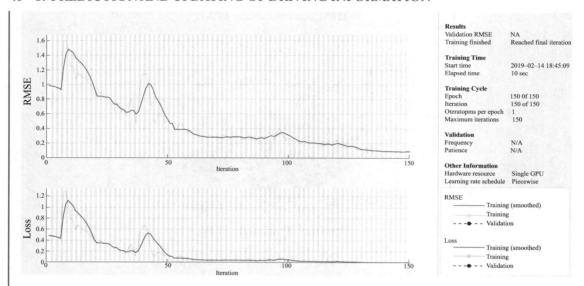

Figure 3.9: Training process of the LSTM network in Matlab.

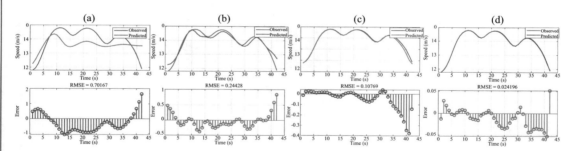

Figure 3.10: Prediction results of LSTM using different training data.

prediction. It can be found that the performance of one-step is better than multi-step because the predicted errors accumulated in the multi-step prediction. Moreover, the accuracy of (c) is higher than that of (a). This can be attributed to the training data, which owes to the driving style and speed interval of the training data in (c) being close to the predicted trajectory. Therefore, when applying the LSTM network for real-time prediction, the experiment data need to be chosen properly, which means the highway driving data are better for time sequence prediction in the highway scenario, and urban driving data are more suitable for urban environments.

In the RL-enabled intelligent energy management system, the predictive EMS is derived by combining the predicted driving trajectories and RL framework. The driving information is forecasted first, and its related TPM is computed based on the prediction results. Then, the differeces between the current TPM and future TPM are measured by the KL divergence rate or

IMN values. When the updating condition is activated, the RL algorithm is applied to generate the new EMS based on the new TPM. By doing this, the EMS can change with the different driving situations and the relevant power split controls are adaptive to various driving environments. The detailed simulation results will be displayed and discussed in the next chapter.

3.4   SUMMARY

Intelligent energy management system means the energy management policy is changeable to adjust to different driving conditions. Future driving information is significant to build this system because the relevant EMS should regulate power distribution according to the future driving cycle and power demand. Regarding the MC model and deep learning method, three potential approaches are introduced to search the future driving information for HEV. To synthesize this information into the RL framework, an online updating algorithm is modified to update the TPM online. This operation enables the transition model in RL so it can change with driving conditions, which implies that the RL-based EMS can always adapt to the driving situation. This characteristic is typically important and difficult for current HEVs on the road. Therefore, the first three chapters aim to build an online and adaptive energy management framework to address this problem. Assuming the calculative efficiency of onboard controllers is fantastic, the RL-enabled intelligent energy management could absolutely improve fuel economy and energy efficiency for HEVs. The next chapter estimates the performance of the proposed system by presenting the simulation and hardware-in-loop (HIL) results.

CHAPTER 4

Evaluation of Intelligent Energy Management System

4.1 BENCHMARK ENERGY MANAGEMENT METHODS

This section aims to evaluate the RL-enabled intelligent energy management system for HEVs since the optimization-based methods have been widely used in the energy management field, and the performance of these approaches has also been estimated in different powertrain architectures and driving situations. This book chooses dynamic programming (DP) and stochastic dynamic programming (SDP) as benchmark technologies to evaluate the RL-based controllers. When the driving information is known in advance, DP method can generate globally optimal results, which can be regarded as a baseline to assess other techniques' effects.

The realization process of SDP includes MDP, policy estimation, and policy improvement, and SDP is suitable to evaluate RL algorithms because their foundation is MC model and TPM.

The estimation process is achieved in three aspects. First, the optimality of the mentioned RL algorithms (Q-learning, Sarsa, Dyna-Q, and Dyna-H) is verified and evaluated by comparing with the benchmark methods. Then, the predictive energy management policy is derived by combining the future driving information and RL algorithms. Finally, to apply the RL-based EMS in real time, the online updating algorithm is integrated into the RL framework to generate and update power split controls online. The future of the intelligent energy management field is also discussed.

4.1.1 DYNAMIC PROGRAMMING-BASED CONTROLLER

Based on Bellman's principle of optimality, DP is able to solve the multi-step horizon optimization problem and guarantee the global optimality by an exhaustive search of all the state variables and control actions. Bellman's principle of optimality states that for the N-steps discrete optimization control problem in Section 2.1.5, if $a(k)(k = 1, 2, \ldots, N)$ is the optimal controls over the whole time horizon, then the truncated sequence $a(k)(k = s + 1, s + 2, \ldots, N)$ is still the optimal controls from time instant $s + 1$ to N.

The cost function in (2.19) is discretized as N-steps control problem

$$J_N(x(1)) = \varphi(x(N)) + \sum_{k=1}^{N-1} L(x(k), a(k), k). \tag{4.1}$$

The optimal cost function J_N^* is minimizing the cost function over the time interval as

$$J_N^*(x(1)) = \min \left\{ \varphi(x(N)) + \sum_{k=1}^{N-1} L(x(k), a(k), k)) \right\}. \tag{4.2}$$

Then, Bellman's principle of optimality can be represented as the Bellman equation

$$J_{N-k}^*(x(k)) = \min \left\{ J_{N-(k+1)}^*(x(k+1)) + L(x(k), a(k), k) \right\}. \tag{4.3}$$

From (4.3), the optimal control policy $\pi^* = \{a^*(1), a^*(2), \ldots, a^*(N)\}$ can be obtained through a backward iteration process. Then, a forward recursive process can be used to calculate the relevant optimal state variables $\{x^*(1), x^*(2), \ldots, x*(N)\}$. The computational sequence is shown in Fig. 4.1. In this book, the DP algorithm is conducted in Matlab based on the toolbox introduced in [97].

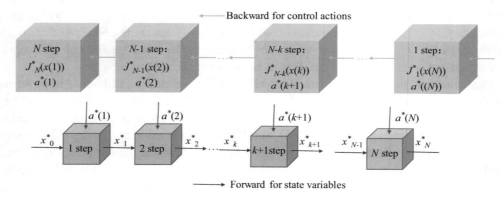

Figure 4.1: Computational sequence of DP algorithm for control actions and state variables.

4.1.2 STOCHASTIC DYNAMIC PROGRAMMING-BASED CONTROLLER

Also, SDP can be applied to formulate the optimal control policy for the energy management problem [98]. The cost function followed by a special policy π is described as

$$J_\pi(t) = L(x(t), a(t), t) + \tau \sum_{t+1} J_\pi(t+1), \tag{4.4}$$

wherein $J_\pi(t)$ means the expected cost related to the selected time point. $\tau \in [0, 1]$ is a factor to discount the future cost and guarantee its convergence.

A policy iteration is usually employed to achieve the SDP algorithm, which consists of two main procedures. These two processes focus on finding a policy and optimizing it, so they are called a policy evaluation and policy improvement. At the beginning of policy evaluation, a random policy π_0 is decided and the corresponding cost $J_{\pi 0}(t)$ is determined in (4.4). Then, the Bellman equation is utilized to generate a new policy as

$$\pi_1 = \arg\min[J_{\pi 0}(t)]. \tag{4.5}$$

In the policy improvement step, the cost function would follow the new control policy and get the relevant expected cost. To find the optimal control policy, these two steps should repeat until the cost function converges at a defined tolerance level. In this book, the SDP algorithm is realized in Matlab using the MDP toolbox in [83].

4.2 OPTIMALITY OF RL-BASED ENERGY MANAGEMENT

4.2.1 EVALUATION OF Q-LEARNING AND SARSA

The studied powertrain in this section is a series hybrid tracked vehicle, as depicted in Fig. 4.2. Its configuration parameters are introduced in Table 2.1. To demonstrate the optimality of the Q-learning algorithm, the SDP-based and RL-based EMSs are compared on a reference driving schedule, which is depicted in Fig. 4.3.

The SOC trajectories and working area of the engine are displayed in Fig. 4.4. Due to the charge sustenance restraint in Equation (2.28), the initial and final SOC values are nearly the same. However, the trend of SOC in RL and SDP is not the same, which is caused by the working points of the engine. In RL case, the engine working points often locate in the lower fuel consumption area, and thus the related fuel consumption would be lower. To eliminate the differences in fuel consumption caused by differences of SOC final values, a SOC-correction

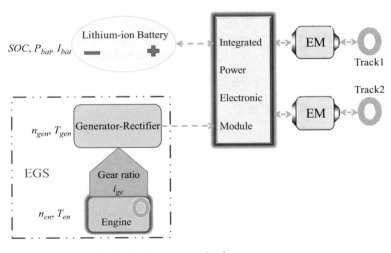

Figure 4.2: The studied series hybrid electric tracked powertrain.

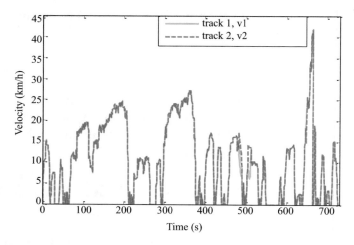

Figure 4.3: Simulation driving schedule for comparison of RL and SDP.

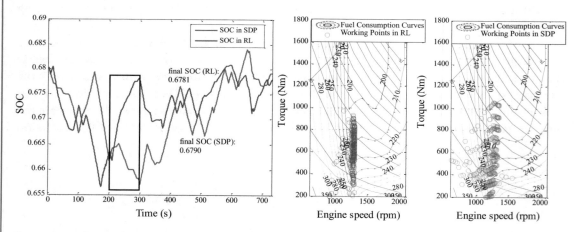

Figure 4.4: Comparison of simulation results between RL and SDP.

method [99] is often adopted to compensate. Table 4.1 gives the fuel consumption of these two methods, in which it is easy to find that Q-learning is 2.4% lower than that of SDP.

The results of power distribution can directly reflect the influences of control actions. Figure 4.5 illustrates the battery and engine power in these two cases. It is obvious that the power distribution is different in some places, which will result in different fuel consumption. For example, the black ellipse in Fig. 4.5 highlights the differences in power distribution between SDP and Q-learning. The power split controls would further affect the SOC variation in battery,

Table 4.1: Tuned fuel consumption in RL and SDP

Methods	Fuel Economy (g)	Relative Increase (%)
Q-learning algorithm	1907	—
SDP approach	1952	2.4

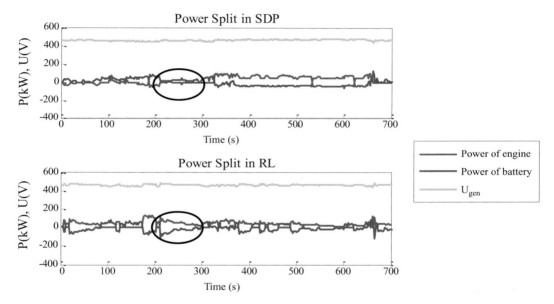

Figure 4.5: Power distribution in two compared approaches.

as the black rectangle shows in Fig. 4.4. These differences will be reflected as the differences of cost function, which means the fuel economy in Table 4.1 is different.

To further estimate the adaptability of RL on another driving schedule, RL and SDP are compared again on a real-time collected cycle (see Fig. 4.6). The relevant simulation results including SOC curves, engine working points, power split, and fuel consumption are shown as below. Overall, the Q-learning-based controls could achieve 1.93% lower fuel consumption than SDP. Based on the advantages of RL in these two driving schedules, the optimality of Q-learning has been proved.

After assessing the optimality of Q-learning algorithm in energy management, a comparison analysis of Q-learning algorithm and Sarsa is conducted on the same hybrid tracked powertrain and driving cycle (that one shown in Fig. 4.6). To represent the convergence rate, the mean discrepancy of these two algorithms at $v = 25$ km/h is displayed in Fig. 4.9. The mean discrepancy indicates the deviation of action-value Q-table per 100 iterations. In Fig. 4.9, the

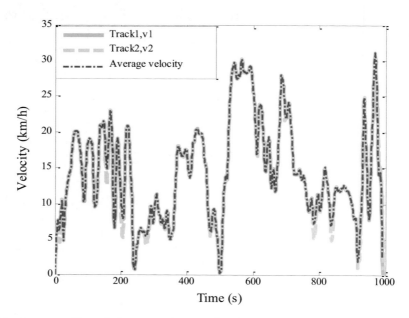

Figure 4.6: Real-time collected driving schedule for Q-learning evaluation.

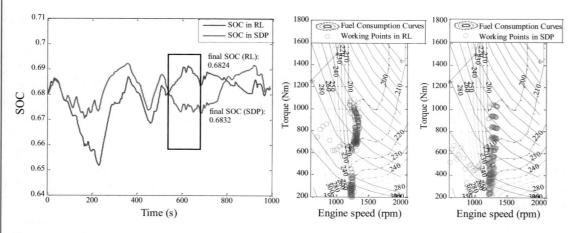

Figure 4.7: Real-time collected driving schedule for Q-learning evaluation.

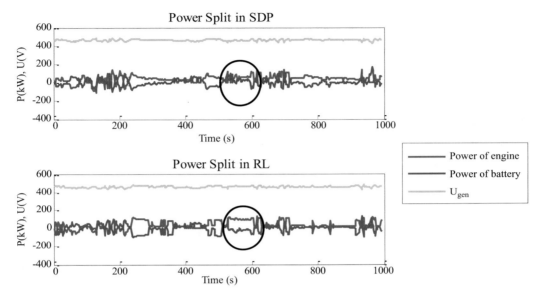

Figure 4.8: Power split situation for the second simulation driving cycle.

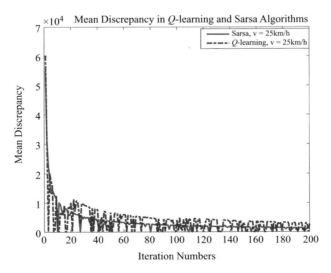

Figure 4.9: Mean discrepancy of action-value Q-table in Q-learning and Sarsa.

Table 4.2: Fuel consumption comparison in RL and SDP for real-time collected cycle

Approaches	Fuel Economy (g)	Relative Increase (%)
Q-learning algorithm	2795	—
SDP method	2849	1.93

Sarsa decreases faster than Q-learning, which means it has a better convergence rate. This is contributed by the ε-greedy policy for the next action selection.

The related simulation results (SOC trajectories, engine working area, and power distribution) between Q-learning and Sarsa are described in Figs. 4.10 and 4.11. At first, the TPM is not mature, and the SOC curves are close. Then, the differences between these two cases are magnified due to the selection rule of the next control action. Q-learning prefers to choose actions with maximum probability and Sarsa would search more probabilities under the ε-greedy policy. As a result, the power distribution between the engine and battery make a difference in these two cases.

Table 4.3 illustrates the fuel consumption and computation time in these two cases. Through the compared analysis, the positives and negatives of these two algorithms are apparent. Sarsa could achieve better fuel economy, but, it would cost more time. Conversely, the performance of Q-learning is worse but can save time for calculation. Therefore, Q-learning is suitable for online application and Sarsa is appropriate for offline optimization.

Finally, the better performance algorithm, Sarsa, is compared with two benchmark methods to further validate optimality. Also, the terminal values of SOC are nearly the same as their initial ones because of the charge sustenance constraint. The fuel consumption in Table 4.4 implies the fuel economy of Sarsa is better than SDP and is extremely close to the DP. Since there are two steps in SDP and each step would consume more time to reach convergence, the computation time of SDP is longer than DP and Sarsa. Overall, this section demonstrates that the RL algorithms could generate sub-optimal control results, which are close to the globally optimal ones and better than some of the optimization-based methods. Thus, RL is promising to be applied in real-time situations in the following sections.

4.2.2   EVALUATION OF DYNA-Q AND DYNA-H

As Dyna-Q and Dyna-H are the extensions of the Q-learning algorithm by synthesizing the planning and learning problems together, this section discusses the performance of these three RL methods. In the cost function, only fuel consumption is intended to be optimized, which indicates the final SOC could be an arbitrary value within the allowable range. In Table 2.4, the planning step implies the degree of exploration at each time instant and would affect the control performance and computation time. Hence, the influence of planning steps in Dyna-H algorithm (steps 8–13 in Table 2.4) on search speed is first assessed. The numbers of the

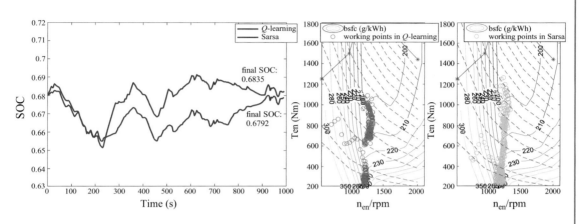

Figure 4.10: SOC in battery and working points in engine of Q-learning and Sarsa.

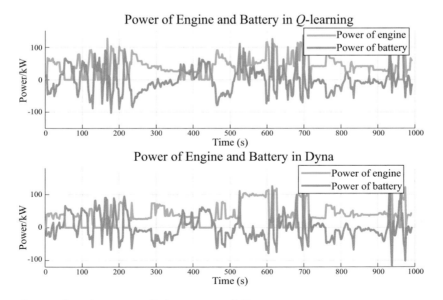

Figure 4.11: Power distribution in Q-learning and Sarsa.

Table 4.3: Fuel consumption and calculative time of Q-learning and Sarsa

Method	Fuel Cost (g)	Relative Increase (%)
Q-learning	2896	1.72
Sarsa	2847	–
Method	**Q-learning Algorithm**	**Sarsa Algorithm**
Time[a] (h)	3	7

Table 4.4: A comparison analysis of DP, Sarsa and SDP

Methods	Consumed Fuel (g)	Relative Increase (%)	
DP	2847	—	
Sarsa	2853	0.21	
SDP	2925	2.74	
Approaches	**DP**	**Sarsa**	**SDP**
Time[a] (h)	2	7	12

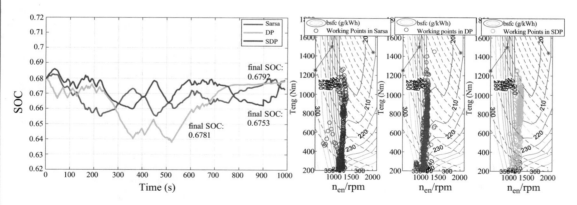

Figure 4.12: Engine operation area and SOC trajectories of Sarsa, SDP and DP.

planning steps are defined as $N = 15, 20, 30$ (step 8 in Table 2.4). Two standard driving cycles named NEDC and UDDS are considered as the simulation cycles, as depicted in Fig. 4.13.

The multiple EMSs derived from the Dyna-H algorithm with different planning steps are implemented in the above two driving cycles. The SOC trajectories in different control cases are displayed in Fig. 4.14. It is easy to observe that the SOC evolutions are different in these two cases. This means the number of planning steps will affect the selection of control actions, and

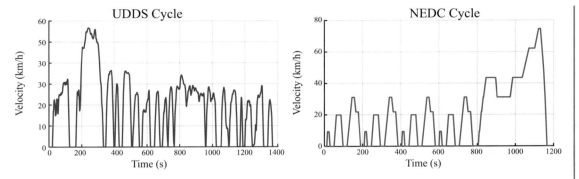

Figure 4.13: UDDS and NEDC for assessment of planning step in Dyna-H.

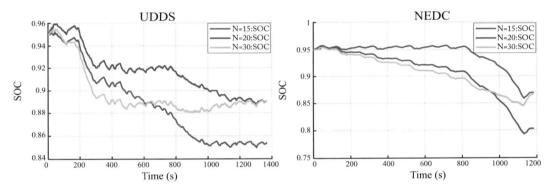

Figure 4.14: SOC variation for different planning steps in two driving cycles.

further influence the power split and cost function. Thus, the appropriate size of the planning step needs to be decided in Dyna-H algorithm.

As the planning step would determine different control actions, it is reasonable to influence the learning rate of the RL method. Figure 4.15 describes the representation of learning rate in Dyna-H algorithm with different planning steps, which is the iteration steps in one episode. An analogous observation that can be found in these two cases is that the learning rate decreases with the planning steps. This is because more trials will be executed in the larger planning step case. However, the computation time will also increase. Thus, it is significant to decide the proper planning step depend on the performance in cost function.

From Fig. 4.16, it is important to see that the larger planning step would not always result in lower fuel consumption. It can be explained in that more explorations may not obtain better control actions because the best one may have been found via ε-greedy policy. Overall, in Dyna-H algorithm, the planning step should be decided appropriately to balance the control objective and learning speed.

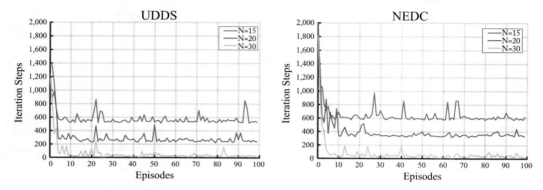

Figure 4.15: Iteration steps in each episode for different planning step cases.

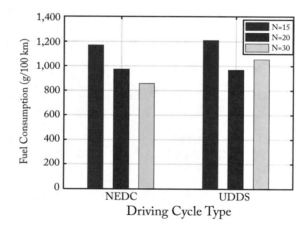

Figure 4.16: Fuel cost for different Dyna-H-based controls in two driving cycles.

After analyzing the significance of the planning step in Dyna-H algorithm, a comparison among Q-learning, Dyna-Q, and Dyna-H is conducted now. The experiment driving cycle is called NYC-HEV, which is taken from software Autonomy. Also, the Dyna-H algorithm chooses multiple planning steps to validate its control performance. In order to guarantee an equitable comparison, the regular parameters in RL framework are the same in these three methods. Five SOC trajectories related to five control cases are given in Fig. 4.17. From the zoom-in figures, it is easy to see these SOC curves are not the same. Also, even in the same Dyna-H algorithm, different planning steps would lead to various SOC trajectories.

Furthermore, to evaluate these three RL approaches on different driving cycles, more driving cycles provided by the software Autonomie are used to validate the aforementioned five control methods. The relevant fuel costs are summarized in Table 4.5, in which the fuel

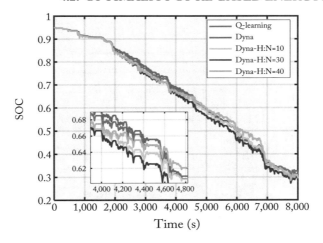

Figure 4.17: SOC trajectories of three RL methods for the same driving cycle.

Table 4.5: Equivalent fuel cost of different control strategies on multiple driving cycles

Cycles / Controls#	REV2	ARB	EPA	NYC-HEV	WLTC
Q-learning	553.5	895.4	1055.5	4967.9*	986.4
Dyna	549.4	842.6	1017.7	4829.8	904.3
Dyna-H: N=10	537.1	817.2	979.9	4724.3	801.5&
Dyna-H: N=30	536.3	787.4	825.2&	4653.5&	839.2
Dyna-H: N=40	531.4&	747.2&	874.1	4761.4	878.1
DP	512.5	711.3	768.1	4475.7	764.8

& Lowest fuel consumption in five learning control policies.
* Equivalent fuel consumption (g/100 km).
# A 2.90 GHz microprocessor with 7.83 GB RAM was used.

consumption has been handled by the SOC-correction method. Compared with the baseline DP method, Dyna-H should achieve better fuel economy than Q-learning and Dyna-Q algorithms. Moreover, a proper planning step size would improve the performance even using the same Dyna-H algorithm. It is usually realized by a cyclic iteration process (added before step 8 in Table 2.4) by considering the cost function and learning rate together.

To represent the search speed of the three RL methods, the following figure describes the iteration steps of each episode. After about 10 episodes, the Dyna-H algorithm could use lower iteration steps to find the optimal controls. It represents that Dyna-H algorithm could decrease

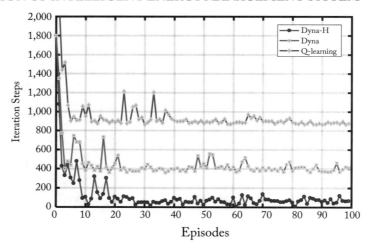

Figure 4.18: Iteration steps of each episode of three RL methods.

the convergence time by incorporating the heuristic function. Thus, this method is promising for application in real-time experiments and verification.

Finally, since the Dyna-H could achieve sub-optimal results by comparing with DP, Table 4.6 lists the computation time of Dyna-H and DP. The compared simulation cycles are the same as those in Table 4.5, and the relevant calculation time indicates the Dyna-H has the potential to be applied in real time. In this book, a reference concept is presented to search the suitable planning step by a periodic iteration. First, we should decide the permitted calculation time depending on the practical problems. Then, this time is regarded as a constraint to search the planning step. The planning step achieves the best performance and its calculative time would not exceed the constraint, meaning it could be chosen as the most appropriate one.

Table 4.6: Comparison of calculative time in DP and Dyna-H

Cycles / Controls[#]	NYC-HEV	WLTC	REV2	EPA
Dyna-H: N=10	102.4[*]	38.3	28.5	49.8
Dyna-H: N=30	116.5	44.9	32.3	57.9
Dyna-H: N=40	123.8	47.5	35.7	64.6
DP	7139	3513	2779	3806

[*] Computational times (s).
[#] A 2.90 GHz microprocessor with 7.83 GB RAM was used.

4.3   RL-BASED PREDICTIVE ENERGY MANAGEMENT

In this section the importance of future driving information in energy management is discussed and analyzed. It also aims to derive and illuminate the predictive energy management policy based on NND, FCG, and RL algorithm. First, the prediction performance for vehicle velocity using NND and FCG is explained. The evaluation rule is also the RMSE in Equation (3.26). For practical application purposes, the simulation cycles (named A and B) are collected from the real-vehicle experiment, as shown in Fig. 4.19.

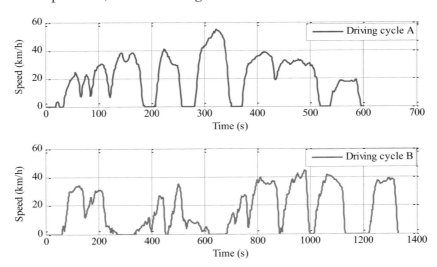

Figure 4.19: Real-vehicle cycles for predictive EMS evaluation.

The predicted one-step results are displayed in Fig. 4.20. As highlighted in purple rectangles and orange ellipses, it is easy to find that the accuracy of FCG is higher than that of NND. The relevant RMSE values are A = 4.103, B = 2.071 for FCG, and A = 8.8697, B = 3.3573 for NND. Also, the 10-step speed prediction is shown in Fig. 4.20 for the same two cycles. From the zoom-in figures and the highlighted spots, it can be found that the performance of FCG is better than NND. The RMSE is A = 3.626, B = 3.516 for FCG, which is lower than that of NND (A = 6.071, B = 4.866).

To derive the predictive EMS, the future velocity information is used to compute to TPM via (2.29); this TPM is regarded as the transition model in the RL framework. Thus, the future driving information is easily embedded into the RL algorithm to obtain the power split controls. The DP-based results and the case without prediction information are selected as two baseline methods. These four control policies are implemented on the other driving cycle. The related SOC curves and power distribution are given in Fig. 4.22, wherein the final SOC value could come back to the initial level. Owing to this fact, the engine would provide all the energy for vehicle running. It can be discerned that SOC in FCG is close to DP and differs from those

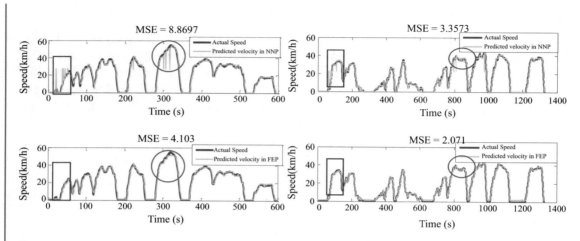

Figure 4.20: One-step speed prediction using NND and FCG.

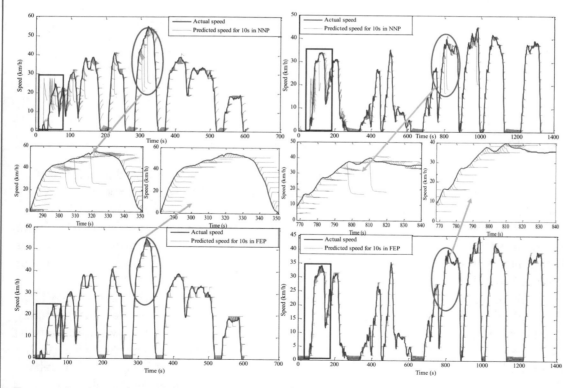

Figure 4.21: 10-step speed prediction using NND and FCG.

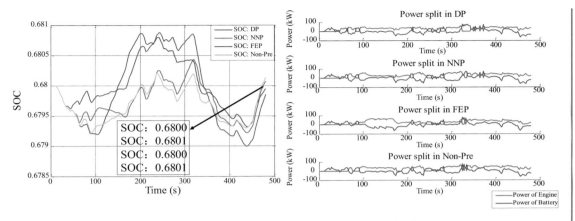

Figure 4.22: SOC variation and power split in four compared methods.

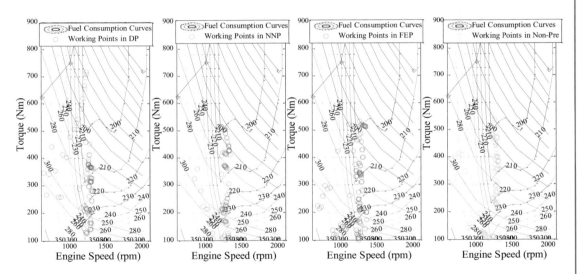

Figure 4.23: Engine working region in DP, predictive and non-predictive cases.

in NND and non-prediction cases. An analogous result in the power split trajectory can also be found.

The working points of the engine in Fig. 4.23 could reflect the fuel cost level. Compared with the non-predictive control, the DP and predictive controls work more effectively in the lower fuel-consumption region, which implies that the prediction speed could help improve fuel economy. To further the fuel economy and real-time potential, Tables 4.7 and 4.8 labeled the fuel consumption and cost time for these four methods. DP achieves the best fuel performance, but, its computation time is the longest. FCG is close to DP and better than NND, and their

Table 4.7: Fuel economy representation in four methods

Algorithms	Fuel Consumption (g)	Relative Increase (%)
Non-Pre	196	13.95
NND	188	9.3
FCG	179	4.07
DP	172	—

Table 4.8: Computation time in DP, NND, FCG and non-predictive cases

Algorithms	Time[a] (min)	Relative increase (%)
DP	8.21	104.23
FCG	5.65	41.27
NND	4.58	13.98
Non-Pre	4.02	—

calculation time is in the opposite situation. It should be noted that the calculative time of predictive EMSs implies that they can be used in the online application.

Finally, an HIL experiment is constructed to mimic the real-time application of predictive EMS. The baseline is chosen as the common rule-based controls, wherein predefined logics are utilized to switch modes among pure electric, hybrid, and charging. An example of hybrid mode and the experimental bench is described in Fig. 4.24. Two main parts are included in the HIL bench, which are MotoTron for control policies storage and RT-Lab for powertrain modeling.

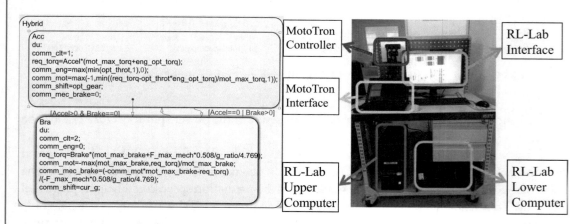

Figure 4.24: HIL experimental bench and hybrid mode in rule-based control.

The experiment results of HIL are showcased in Fig. 4.25. The performance of proposed predictive EMS is far better than that of rule-based control. The fuel consumption of the former is 17.54% lower than that of the latter. Thus, we could conclude that the proposed predictive control strategy is more fuel-saving when compared with the common onboard rule-based controls while possessing real-time applicability.

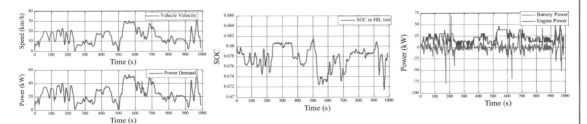

Figure 4.25: Simulation results in HIL experiment for a special driving cycle.

4.4   EVALUATION OF REAL-TIME ENERGY MANAGEMENT

It has been demonstrated that the RL-enabled EMS could achieve fantastic control performance and is also usable for real-time application. In the previous discussion, the simulation driving cycle is assumed to be known in advance. In this section, we discuss the online application of RL-enabled intelligent EMS in two situations. First, the current driving conditions are compared with the near past driving conditions via KL divergence rate or IMN (introduced in Section 2.2). The online updating algorithm introduced in Section 3.2 is employed to update the TPM (or transition model in RL) in real time. Second, the predictive power demand, online updating algorithm, and RL methods are integrated. The differences between current and future TPM are recognized, and then the online updating algorithm and RL methods are combined to derive the new EMS to accommodate the future driving conditions.

Taking KL divergence rate as an example, its threshold value would influence the updating frequency of TPM and thus affect the control performance of the presented online EMS. The powertrain is also the series hybrid tracked vehicle and the representative driving cycles (named driving cycle 1 and 2 and given in Fig. 4.26. When the KL values surpass their threshold values, the control strategies are triggered to be replaced. Three threshold values are defined as 0.2, 0.4, and 0.6, respectively, and the forgetting factor in these cases is $\gamma = 0.02$.

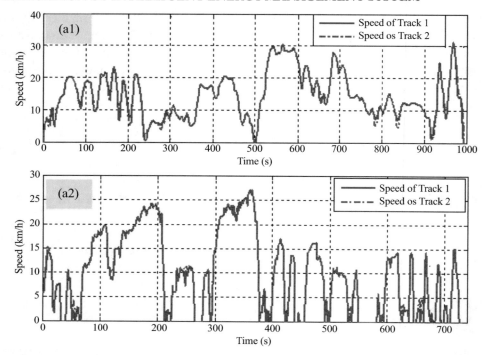

Figure 4.26: Driving cycles to estimate influence of KL divergence rate values: (a1) driving cycle 1; (a2) driving cycle 2.

The SOC trajectories of different KL threshold values in these two simulation driving cycles are illustrated in Fig. 4.27. It is apparent that SOC curves for driving cycle 1 are significantly different but almost the same in cycle 2. That is the result of the updating of TPM; and the updating time instants are shown in Fig. 4.28. The TPMs in Fig. 4.28 are compared per 100 seconds. For different KL threshold values, the updating times of TPM are different. Since the KL threshold values are defined in advance, the updating times of control policies are decided. Figure 4.28 suggests that a lower threshold value would lead to more updating of EMS and thus the improvement of control performance. However, the computation burden is heavier.

Table 4.9 explains the fuel consumption and updating times of TPMs for different KL threshold values. In driving cycle 2, the updating times are almost the same, which implies that the relevant EMSs are the same, and thus the SOC curves in Fig. 4.27 would be nearly the same. Instead, the different updating times of TPM and control strategies in driving cycle 1 result in different performance of fuel economy and SOC trajectories. To consider performance and computation efficiency together, the proper KL threshold value is defined as 0.4 in the online control.

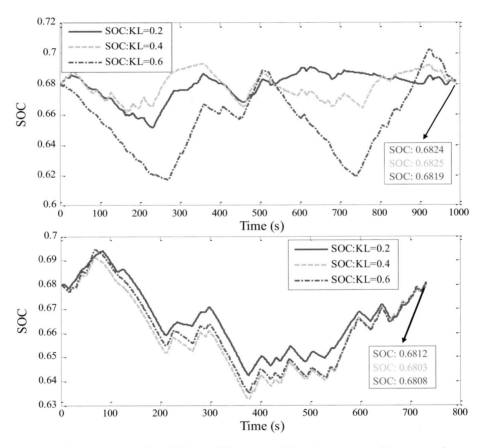

Figure 4.27: SOC trajectories for different KL threshold values in two driving cycles.

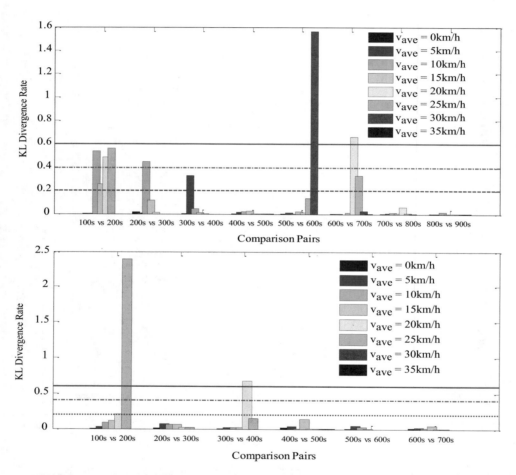

Figure 4.28: Comparison of TPM using KL divergence rate values in two driving cycles.

Table 4.9: Updating times of TPM and fuel consumption in different control cases

KL Threshold Values	Updating Times[1]	Fuel Cost (g)[1]	Updating Times[2]	Fuel Cost (g)[2]
0.2	9	2723	3	1576
0.4	6	2754	2	1583
0.6	2	2786	2	1583
[1]Driving cycle 1; [2] Driving cycle 2.				

Another important parameter in the online updating algorithm is the forgetting factor. The forgetting factor determines the memory depth, which decides the relevance of the previous and current period driving cycle. To study the influence of the forgetting factor, we defined its value as 0.02, 0.01, and 0.005 and KL threshold values as 0.2, 0.4, and 0.6 to see the fuel economy in different combinations. Figure 4.29 gives the fuel consumptions when the KL threshold value and forgetting factor vary. Note that the smallest forgetting factor does not guarantee the best fuel economy; even the smaller forgetting factor indicates a deeper memory depth because the next period driving cycle does not always have the largest weight for the smallest forgetting factor case. Therefore, proper matching between the KL threshold value and forgetting factor should be decided for online application of RL-enabled EMS.

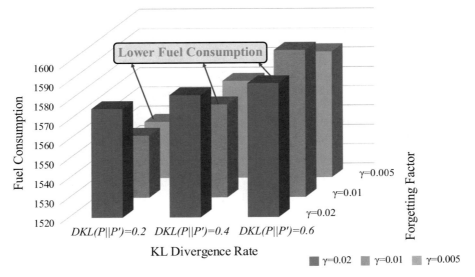

Figure 4.29: Fuel consumption in different combinations of forgetting factor and KL threshold values.

After analyzing the influence of two parameters in online RL-based EMS, a comparison of DP, online, and offline RL-enabled EMSs is executed. In the online EMS, the forgetting factor is $\gamma = 0.01$ and KL threshold value is 0.4. The initial SOC value is 0.68 and the sample time is 1 second. Figure 4.30 shows the SOC and power split trajectories, in which the online results are close to the those of DP. This observation can be contributed by the updating of the control strategy. Figure 4.31 explains the practical updating time instants, at 300 s, 500 s, and 700 s. The RL-enabled EMS is triggered to update to accommodate changed driving conditions.

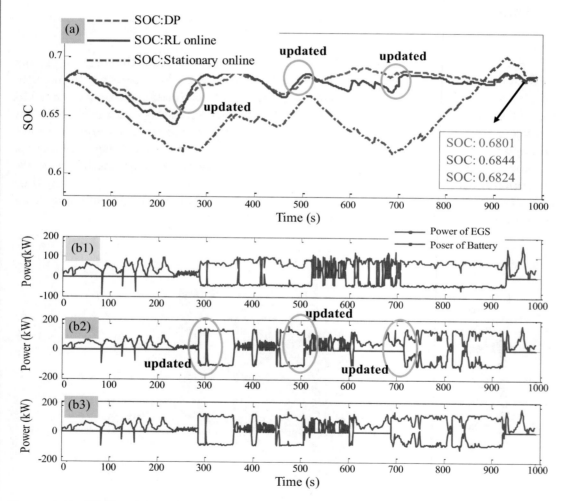

Figure 4.30: SOC and power split trajectories in three control cases: (b1) DP, (b2) online, and (b3) offline controls.

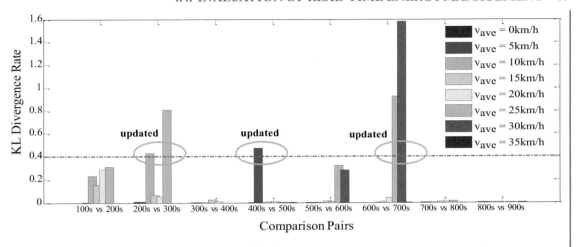

Figure 4.31: Comparison of TPM using KL divergence rate values in three control cases.

Similarly, the working points of the engine and the resulting fuel consumption are displayed in Fig. 4.32 and Table 4.10, respectively. Compared with the offline control, the working points in online control and DP are located in a lower consumption area. The number in Table 4.10 reflects that the fuel economy of online control is close to that of DP and 1.53% lower than that under offline control. Moreover, the consumed time in Table 4.11 shows the online control could save time compared with the other two methods, which makes online optimization feasible.

Table 4.10: Consumed fuel in three compared methods

Methods	Fuel Economy (g)	Relative Increase (%)
DP	2708	-
Online control	2753	1.68
Offline control	2795	3.21

Table 4.11: Time cost in three compared methods

Methods	Time Cost^a (h)	Relative Increase (%)
Online control	1.2	-
DP	2	66.67
Offline control	4	233.33
^{a} A 2.4 GHz microprocessor with 12 GB RAM was used.		

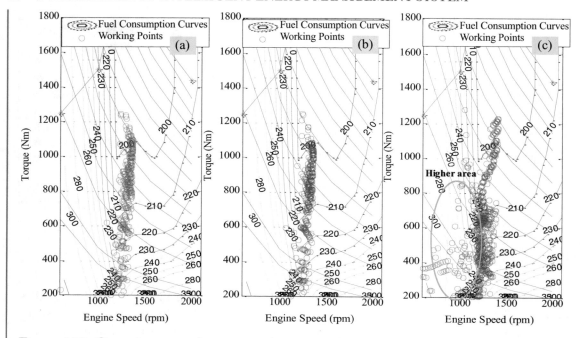

Figure 4.32: Operation area of engine in three control strategies, (a) DP, (b) online, and (c) offline.

Finally, in the first simulation situation, to verify the robustness of the proposed online RL-enabled EMSs, the previous three mentioned controls are implemented for the other two driving cycles (named C and D). The fuel consumption is shown in Table 4.12, wherein a similar conclusion is obtained that the performance of online control is close to DP and better than offline control. These findings validate the robust performance of the proposed online EMS.

Table 4.12: Robustness validation of the online control strategy on two other cycles

Driving Cycles	Controls	Consumed Fuel (g)	Final *SOC*	Relative Increase (%)
Cycle C	DP	1455	0.6803	-
	Online control	1471	0.6815	1.1
	Offline control	1532	0.6811	5.29
Cycle D	DP	3518	0.6807	-
	Online control	3545	0.6842	0.77
	Offline control	3608	0.6821	2.56

In the second simulation, future driving information is considered in the online control strategy. Taking the power demand as an example, the one-step and 10-step predicted power demand for two driving cycles are shown in Fig. 4.33. The KL threshold value and forgetting factor are defined as 0.4 and 0.01, respectively. Also, three control strategies (DP, predictive online EMS, and primary RL-based EMS) are compared on the same cycle, as depicted in Fig. 4.34. The corresponding SOC and power split curves are given in Fig. 4.35. As can be seen, the SOC in predictive RL is closer to that of DP than the primary RL. In the power split curves, an analogous conclusion can be found. This improvement can be ascribed to the online updating of EMS guided by future power demand information.

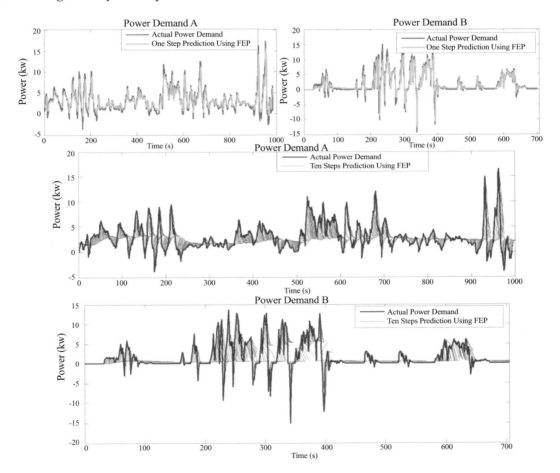

Figure 4.33: Power demand prediction with one-step and 10-step dimension.

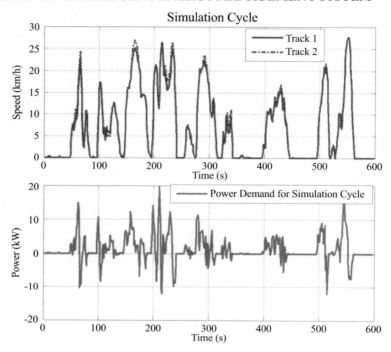

Figure 4.34: Simulation driving cycle for the comparison of three control policies.

Figure 4.36 also describes the updating time instants in predictive online control, which are 200 and 300 s. Owing to this updating, the onboard energy management controller could adjust the power split controls accordingly, and thus the related EMS could adapt to the future driving conditions. Finally, the fuel economy and consumed time are also given in Tables 4.13 and 4.14. It is easy to recognize that predictive online control is lower than the common RL by 4.11%, and its computation time is the shortest in these three cases. By integrating the predictive driving information and online updating algorithm into the RL framework, the EMS could be more intelligent to accommodate the various driving conditions. Therefore, the proposed RL-enabled energy management system has potential to be applied in the real vehicle to improve fuel economy and achieve other optimization control goals.

Table 4.13: Fuel consumption in DP, predictive online RL and common RL

Control Methods	Fuel Cost (g)	Relative Difference (%)
DP	260.3	-
Common RL	277.5	6.61
Predictive Online RL	266.8	2.5

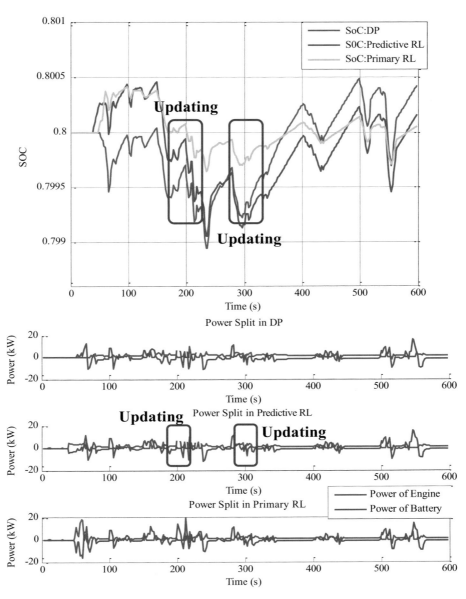

Figure 4.35: SOC and power distribution results in three controls.

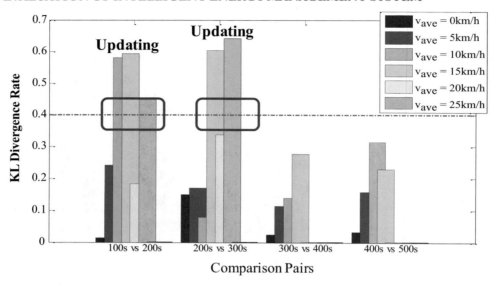

Figure 4.36: Computation and comparison of KL divergence rate per 100 s.

Table 4.14: Consumed time in DP, predictive online RL, and common RL

Control Methods	Consumed Time[a] (h)	Relative Difference (%)
Predictive Online RL	1.42	-
Common RL	4.65	227.46
DP	2.58	81.69

4.5   SUMMARY

Comprehensive evaluation of RL-enabled EMSs is conducted in this chapter. The global optimal DP-based controls and sub-optimal SDP-based control are treated as the benchmark methods. First, the optimality of the different RL algorithms, Q-learning, Sarsa, Dyna-Q, and Dyna-H, are estimated. The balance of computational efficiency and control performance is necessary to be decided when choosing the RL algorithms for practical problems. Then, the future driving information is alone to be considered in RL-based controls. It is easy to see that future driving information could help improve fuel economy and calculative efficiency. Finally, we combined the future power demand, online updating algorithm, and RL framework to construct the intelligent energy management system. The advantages of this system are explained and demonstrated. It is concluded that this RL-enabled intelligent energy management system has potential to be applied in real vehicles for online applications. This action could further improve the ability of HEVs in fuel saving and pollution reduction.

CHAPTER 5

Conclusion

One of the future directions for research is applying more efficient artificial intelligence (AI) approaches in the energy management field of HEV. The theoretical feasibility could be validated by simulation, and the practical implementation should be conducted in real-vehicle evaluation. Additional major work in the future is to access and improve energy management strategies in the intelligent transportation environment. Since the traffic information can be acquired, how to attune the strategies to other vehicles' and infrastructures' behaviors should be further addressed.

This section discusses future perspectives and trends of the RL-based energy management of HEV. These prospects consist of four circumstances: (1) the novel and efficient RL algorithms are going to be applied in this field; (2) energy management integrates with intelligent transportation systems (ITS) to construct a smart city or smart grid; (3) optimization control objectives become more and more comprehensive and complicated; (4) distributed or multi-agent RL system for cooperative learning between vehicles occurs in the connected vehicle environment.

Novel RL Algorithms

Fast development of calculative capacity enables the neoteric algorithms to be used in the energy management field of HEVs. Different kinds of deep learning algorithms can be exploited to classify, train, and learn the massive scale of data. For example, deep belief networks (DBN), stacked auto-encoders (SAE) and recurrent neural network (RNN) show promise for being used to learn the special model or table from the generous data. Then, double Q-learning, speedy Q-learning, and deep deterministic policy gradient algorithms are able to formulate the optimal policy based on the trained model or table. With the help of cloudy control and management, these methods are practical and useful for the real-time operations of HEV.

Furthermore, other RL algorithms are proved to be superior to the traditional Q-learning or Dyna, such as k-nearest neighbors temporal difference (kNN-TD(λ)) and Dyna-H. In the former algorithm, a kNN method could represent probabilistic characteristics of the state variables, and TD back-propagation is used to learn the control actions. Dyna-H is a model-free online algorithm, which adds a heuristic planning strategy into a Dyna agent to choose the optimal controls more efficiently. Finally, inverse reinforcement learning (IRL) is also suitable for energy management problems, especially when the control objectives are unknown. This method can search for the proper reward signal through trial and error and learn from existing experience.

Energy Management in Intelligent Transportation System

Information from emerging intelligent transportation system (ITS) technologies (e.g., vehicle to everything (V2X) communication) provides great assistance for the improvement of energy management, such as real-time trip information, specific traffic situations, cloudy prediction, and weather conditions. For example, future trip information can be learned and forecasted from historical driving data. Based on this information, the energy management strategy can be more adaptive and robust to dynamic driving conditions, especially for (P)HEV with stationary routes. The current onboard devices have the ability to get real-time traffic situations by wireless communication, global position system (GPS), and geographical information systems (GIS). The obtained information can regulate online control strategy with advanced computation methods.

Furthermore, future vehicle velocity and power demand are also powerful information to influence the power split controls of an HEV. Developing reliable algorithms may be utilized to acquire these data in a cloud platform or vehicle-to-everything (V2X) communication. The data processing can also be executed online, and then the achieved controls are feasible for a group of vehicles. Finally, weather conditions are essential factors for driver behavior, fuel cost, and electricity cost. The wind direction and temperature may affect the aerodynamic or rolling resistance, and the weather may affect the driving style of different drivers. How to adjust the energy management strategy according to weather information is an open question.

Combination of Multiple Objectives

A transition from one common objective (fuel cost) to multiple goals is another research interest in the energy management field. These objectives include greenhouse gas emissions, state of health (SOH) in battery, safety, and comfort capability, user convenience, and powertrain mobility. Eco-driving is a promising pattern to reduce the usage of ICE to lower emissions. Battery health is a critical parameter to limit the driving mileage of a vehicle with electric power. Drivability for safety and comfort are significant for current vehicles with human drivers. Researchers need to strive for these representative objectives in future energy management research. More important, how to handle the enormous computational overhead provided by the multiple control objectives also needs to be solved in the future.

Cooperative Learning in a Connected Environment

Vehicle automation is another research hotspot in the automobile industry. In the future, a group of HEVs could communicate with each other and their driving behaviors may affect each other. In this connected and networked environment, the central controller should not only consider improving one vehicle's energy efficiency. The energy management controller should realize each vehicle's control objective with taking other vehicles' influences into account. This goal should be achieved by advanced learning method. For example, the asynchronous variants of standard RL algorithms are proposed in [100]. This concept trains deep neural network through asynchronous gradient descent and can not only shorten computation time but also realize parallel calculation. By doing this, different energy management problems can be solved in a connected environment by taking into account other's driving behaviors.

References

[1] W. Liu, *Introduction to Hybrid Vehicle System Modeling and Control*, Wiley, Hoboken, 2013. DOI: 10.1002/9781118407400. 1

[2] L. Guzzella and A. Sciarretta, *Vehicle Propulsion Systems: Introduction to Modeling and Optimization*, Springer, Berlin, 2013. DOI: 10.1007/978-3-642-35913-2. 1

[3] T. Liu, B. Wang, and C. Yang, Online Markov chain-based energy management for a hybrid tracked vehicle with speedy Q-learning, *Energy*, vol. 160, pp. 544–555, 2018. DOI: 10.1016/j.energy.2018.07.022. 1

[4] C. M. Martinez and D. P. Cao, *iHorizon-Enabled Energy Management for Electrified Vehicles*, Butterworth–Heinemann, Elsevier, 2018. DOI: 10.1016/C2017-0-02869-0. 1, 9

[5] X. Tang, D. Zhang, T. Liu, A. Khajepour, H. Yu, and H. Wang, Research on the energy control of a dual-motor hybrid vehicle during engine start-stop process, *Energy*, vol. 166, pp. 1181–1193, 2019. DOI: 10.1016/j.energy.2018.10.130. 1

[6] Y. Zou, T. Liu, F. Sun, and H. Peng, Comparative study of dynamic programming and Pontryagin's minimum principle on energy management for a parallel hybrid electric vehicle, *Energies*, vol. 6, no. 4, pp. 2305–2318, 2013. DOI: 10.3390/en6042305. 1

[7] M. Pourabdollah, B. Egardt, N. Murgovski, and A. Grauers, Convex optimization methods for powertrain sizing of electrified vehicles by using different levels of modeling details, *IEEE Transactions on Vehicular Technology*, vol. 67, no. 3, pp. 1881–1893, 2018. DOI: 10.1109/tvt.2017.2767201. 1

[8] M. Kim and H. Peng, Power management and design optimization of fuel cell/battery hybrid vehicles, *Journal of Power Sources*, vol. 165, no. 2, pp. 819–832, 2007. DOI: 10.1016/j.jpowsour.2006.12.038. 1

[9] R. Johri and Z. Filipi, Optimal energy management of a series hybrid vehicle with combined fuel economy and low-emission objectives, in *Proc. Institution of Mechanical Engineers, Part D: Journal of Automobile Engineering*, vol. 228, no. 12, pp. 1424–1439, 2014. DOI: 10.1177/0954407014522444. 1, 9

[10] A. Castaings, W. Lhomme, R. Trigui, and A. Bouscayrol, Comparison of energy management strategies of a battery/supercapacitors system for electric vehicle under real-time constraints, *Applied Energy*, vol. 163, pp. 190–200, 2016. DOI: 10.1016/j.apenergy.2015.11.020. 2

[11] S. Althaher, P. Mancarella, and J. Mutale, Automated demand response from home energy management system under dynamic pricing and power and comfort constraints, *IEEE Transactions on Smart Grid*, vol. 6, no. 4, pp. 1874–1883, 2015. DOI: 10.1109/tsg.2014.2388357. 2

[12] S. J. Russell and P. Norvig, Artificial intelligence: A modern approach, *Pearson Education Limited*, Malaysia, 2016. 2

[13] R. S. Sutton and A. G. Barto, *Reinforcement Learning: An Introduction*, MIT Press, 2018. DOI: 10.1109/tnn.1998.712192.

[14] Y. Wu, H. Tan, J. Peng, H. Zhang, and H. He, Deep reinforcement learning of energy management with continuous control strategy and traffic information for a series-parallel plug-in hybrid electric bus, *Applied Energy*, vol. 247, pp. 454–466, 2019. DOI: 10.1016/j.apenergy.2019.04.021.

[15] S. M. Zahraee, M. K. Assadi, and R. Saidur, Application of artificial intelligence methods for hybrid energy system optimization, *Renewable and Sustainable Energy Reviews*, vol. 66, pp. 617–630, 2016. DOI: 10.1016/j.rser.2016.08.028. 2

[16] E. Gibney, Google AI algorithm masters ancient game of Go, *Nature News*, vol. 529, no. 7587, p. 445, 2016. DOI: 10.1038/529445a. 2, 9

[17] V. Mnih, K. Kavukcuoglu, D. Silver, A. Rusu, and J. Veness, Human-level control through deep reinforcement learning, *Nature*, vol. 518, no. 7540, p. 529, 2015. DOI: 10.1038/nature14236. 2

[18] T. K. Das, A. Gosavi, S. Mahadevan, and N. Marchalleck, Solving semi-Markov decision problems using average reward reinforcement learning, *Management Science*, vol. 45, no. 4, pp. 560–574, 1999. DOI: 10.1287/mnsc.45.4.560. 2

[19] G. Wu, X. Zhang, and Z. Dong, Powertrain architectures of electrified vehicles: Review, classification and comparison, *Journal of the Franklin Institute*, vol. 352, no. 2, pp. 425–448, 2015. DOI: 10.1016/j.jfranklin.2014.04.018. 2

[20] K. C. Bayindir, M. A. Gözüküçük, and A. Teke, A comprehensive overview of hybrid electric vehicle: Powertrain configurations, powertrain control techniques and electronic control units, *Energy Conversion and Management*, vol. 52, no. 2, pp. 1305–1313, 2011. DOI: 10.1016/j.enconman.2010.09.028. 4

[21] L. Serrao, S. Onori, and G. Rizzoni, A comparative analysis of energy management strategies for hybrid electric vehicles, *Journal of Dynamic Systems, Measurement, and Control*, vol. 133, no. 3, p. 031012, 2011. DOI: 10.1115/1.4003267. 5, 6

[22] Y. Huang, H. Wang, A. Khajepour, H. He, and J. Ji, Model predictive control power management strategies for HEVs: A review, *Journal of Power Sources*, vol. 341, pp. 91–106, 2017. DOI: 10.1016/j.jpowsour.2016.11.106. 5, 6

[23] Y. Huang, H. Wang, A. Khajepour, B. Li, J. Ji, K. Zhao, and C. Hu, A review of power management strategies and component sizing methods for hybrid vehicles, *Renew Sustain Energy Review*, vol. 96, pp. 132–144, 2018. DOI: 10.1016/j.rser.2018.07.020. 5, 6

[24] A. Panday and H. Bansal, A review of optimal energy management strategies for hybrid electric vehicle, *International Journal of Vehicular Technology*, vol. 2014, 2014. DOI: 10.1155/2014/160510. 5, 6

[25] F. Salmasi, Control strategies for hybrid electric vehicles: Evolution, classification, comparison, and future trends, *IEEE Transactions on Vehicular Technology*, vol. 56, no. 5, pp. 2393–2404, 2007. DOI: 10.1109/tvt.2007.899933. 6

[26] A. Malikopoulos, Supervisory power management control algorithms for hybrid electric vehicles: A survey, *IEEE Transactions on Intelligent Transportation Systems*, vol. 15, no. 5, pp. 1869–1885, 2014. DOI: 10.1109/tits.2014.2309674. 6

[27] P. Zhang, F. Yan, and C. Du, A comprehensive analysis of energy management strategies for hybrid electric vehicles based on bibliometrics, *Renew Sustain Energy Review*, vol. 48, pp. 88–104, 2015. DOI: 10.1016/j.rser.2015.03.093. 5, 6

[28] C. Martinez, X. Hu, D. Cao, E. Velenis, B. Gao, and M. Wellers, Energy management in plug-in hybrid electric vehicles: Recent progress and a connected vehicles perspective, *IEEE Transactions on Vehicular Technology*, vol. 66, no. 6, pp. 4534–4549, 2017. DOI: 10.1109/tvt.2016.2582721. 5, 6

[29] M. Sabri, K. Danapalasingam, and M. Rahmat, A review on hybrid electric vehicles architecture and energy management strategies, *Renew Sustain Energy Review*, vol. 53, pp. 1433–1442, 2016. DOI: 10.1016/j.rser.2015.09.036. 5, 6

[30] W. Enang and C. Bannister, Modelling and control of hybrid electric vehicles (a comprehensive review), *Renew Sustain Energy Review*, vol. 74, pp. 1210–1239, 2017. DOI: 10.1016/j.rser.2017.01.075. 6

[31] C. Samanta, S. Padhy, S. Panigrahi, and B. Panigrahi, Hybrid swarm intelligence methods for energy management in hybrid electric vehicles, *IET Electric Systems in Transportation*, vol. 3, no. 1, pp. 22–29, 2013. DOI: 10.1049/iet-est.2012.0009. 6, 9

[32] C. Sun, F. Sun, and H. He, Investigating adaptive-ECMS with velocity forecast ability for hybrid electric vehicles, *Applied Energy*, vol. 185, pp. 1644–1653, 2017. DOI: 10.1016/j.apenergy.2016.02.026. 6, 9

[33] C. Musardo, G. Rizzoni, Y. Guezennec, and B. Staccia, A-ECMS: An adaptive algorithm for hybrid electric vehicle energy management, *European Journal of Control*, vol. 11, no. 4–5, pp. 509–524, 2005. DOI: 10.1109/cdc.2005.1582424. 6, 9

[34] J. Liu and H. Peng, Modeling and control of a power-split hybrid vehicle, *IEEE Transactions on Control Systems Technology*, vol. 16, no. 6, pp. 1242–1251, 2008. DOI: 10.1109/tcst.2008.919447. 6, 9

[35] M. O'Keefe and T. Markel, Dynamic programming applied to investigate energy management strategies for a plug-in HEV, (No. NREL/CP-540–40376), *National Renewable Energy Laboratory (NREL)*, Golden, CO, 2006. 9

[36] B. Chen, Y. Wu, and H. Tsai, Design and analysis of power management strategy for range extended electric vehicle using dynamic programming, *Applied Energy*, vol. 113, pp. 1764–1774, 2014. DOI: 10.1016/j.apenergy.2013.08.018. 6, 9

[37] C. Xu, A. Al-Mamun, S. Geyer, and H. Fathy, A dynamic programming-based real-time predictive optimal gear shift strategy for conventional heavy-duty vehicles, in *Proc. Annual American Control Conference, (ACC)*, pp. 5528–5535, June 2018. DOI: 10.23919/acc.2018.8430948. 7, 9

[38] J. Fu, S. Song, Z. Fu, and J. Ma, Real-time implementation of optimal control considering gear shifting and engine starting for parallel hybrid electric vehicle based on dynamic programming, *Optimal Control Applications and Methods*, vol. 39, no. 2, pp. 757–773, 2018. DOI: 10.1002/oca.2375. 7, 9

[39] T. Liu, H. Yu, H. Guo, Y. Qin, and Y. Zou, Online energy management for multimode plug-in hybrid electric vehicles, *IEEE Transactions on Industrial Informatics*. DOI: 10.1109/tii.2018.2880897. 7

[40] M. Marzband, F. Azarinejadian, M. Savaghebi, and J. Guerrero, An optimal energy management system for islanded microgrids based on multiperiod artificial bee colony combined with Markov chain, *IEEE Systems Journal*, vol. 11, no. 3, pp. 1712–1722, 2017. DOI: 10.1109/jsyst.2015.2422253. 7, 9

[41] T. Sousa, H. Morais, Z. Vale, P. Faria, and J. Soares, Intelligent energy resource management considering vehicle-to-grid: A simulated annealing approach, *IEEE Transactions on Smart Grid*, vol. 3, no. 1, pp. 535–542, 2012. DOI: 10.1109/tsg.2011.2165303. 7, 9

[42] T. Nüesch, P. Elbert, M. Flankl, C. Onder, and L. Guzzella, Convex optimization for the energy management of hybrid electric vehicles considering engine start and gearshift costs, *Energies*, vol. 7, no. 2, pp. 834–856, 2014. DOI: 10.3390/en7020834. 7, 9

[43] X. Hu, N. Murgovski, L. Johannesson, and B. Egardt, Optimal dimensioning and power management of a fuel cell/battery hybrid bus via convex programming, *IEEE/ASME Transactions on Mechatronics*, vol. 20, no. 1, pp. 457–468, 2015. DOI: 10.1109/tmech.2014.2336264. 7, 9

[44] B. Gao, W. Zhang, Y. Tang, M. Hu, M. Zhu, and H. Zhan, Game-theoretic energy management for residential users with dischargeable plug-in electric vehicles, *Energies*, vol. 7, no. 11, pp. 7499–7518, 2014. DOI: 10.3390/en7117499. 7, 9

[45] C. Dextreit, F. Assadian, I. Kolmanovsky, J. Mahtani, and K. Burnham, Hybrid electric vehicle energy management using game theory, *SAE Technical Paper*, 2008. DOI: 10.4271/2008-01-1317. 7, 9

[46] F. Syed, M. Kuang, M. Smith, S. Okubo, and H. Ying, Fuzzy gain-scheduling proportional—integral control for improving engine power and speed behavior in a hybrid electric vehicle, *IEEE Transactions on Vehicular Technology*, vol. 58, no. 1, pp. 69–84, 2009. DOI: 10.1109/tvt.2008.923690. 7, 9

[47] S. Burch, M. Cuddy, and T. Markel, ADVISOR 2.1 documentation, *National Renewable Energy Laboratory*, 1999. 7, 9

[48] X. Tang, D. Zhang, T. Liu, A. Khajepour, H. Yu, and H. Wang, Research on the energy control of a dual-motor hybrid vehicle during engine start-stop process, *Energy*, vol. 166, pp. 1181–1193, 2019. DOI: 10.1016/j.energy.2018.10.130. 7, 9

[49] X. Zeng and J. Wang, A parallel hybrid electric vehicle energy management strategy using stochastic model predictive control with road grade preview, *IEEE Transactions on Control Systems Technology*, vol. 23, no. 6, pp. 2416–2423, 2015. DOI: 10.1109/tcst.2015.2409235. 7, 9

[50] H. Borhan, A. Vahidi, A. M. Phillips, M. L. Kuang, I. V. Kolmanovsky, and S. Di Cairano, MPC-based energy management of a power-split hybrid electric vehicle, *IEEE Transactions on Control Systems Technology*, vol. 20, no. 3, pp. 593–603, May 2012. DOI: 10.1109/tcst.2011.2134852. 7, 9

[51] B. Sampathnarayanan, L. Serrao, S. Onori, G. Rizzoni, and S. Yurkovich, Model predictive control as an energy management strategy for hybrid electric vehicles, *Proc. ASME Dynamic Systems and Control Conference*, pp. 249–256, American Society of Mechanical Engineers. DOI: 10.1115/dscc2009-2671. 7, 9

[52] Z. Song, J. Hou, H. Hofmann, et al., Sliding-mode and Lyapunov function-based control for battery/supercapacitor hybrid energy storage system used in electric vehicles, *Energy*, vol. 122, pp. 601–612, 2017. DOI: 10.1016/j.energy.2017.01.098. 7, 9

[53] H. Tian, Z. Lu, et al., A length ratio based neural network energy management strategy for online control of plug-in hybrid electric city bus, *Applied Energy*, vol. 177, pp. 71–80, 2016. DOI: 10.1016/j.apenergy.2016.05.086. 7, 9

[54] T. Liu, Y. Zou, D. Liu, and F. Sun, Reinforcement learning of adaptive energy management with transition probability for a hybrid electric tracked vehicle, *IEEE Transactions on Industrial Electronics*, vol. 62, no. 12, 7837–7846, 2015. DOI: 10.1109/tie.2015.2475419. 7, 9

[55] T. Liu, Y. Zou, D. Liu, and F. Sun, F. Reinforcement learning-based energy management strategy for a hybrid electric tracked vehicle, *Energies*, vol. 8, no. 7, pp. 7243–7260, 2015. DOI: 10.3390/en8077243. 7, 9

[56] Y. Zou, T. Liu, D. Liu, and F. Sun, Reinforcement learning-based real-time energy management for a hybrid tracked vehicle, *Applied Energy*, vol. 171, pp. 372–382, 2016. DOI: 10.1016/j.apenergy.2016.03.082. 7, 9, 11, 35

[57] T. Liu, X. Hu, S. E. Li, and D. Cao, Reinforcement learning optimized look-ahead energy management of a parallel hybrid electric vehicle, *IEEE/ASME Transactions on Mechatronics*, vol. 22, no. 4, 1497–1507, 2017. DOI: 10.1109/tmech.2017.2707338. 9

[58] T. Liu and X. Hu, A bi-level control for energy efficiency improvement of a hybrid tracked vehicle, *IEEE Transactions on Industrial Informatics*, vol. 14, no. 4, pp. 1616–1625, 2018. DOI: 10.1109/tii.2018.2797322. 7, 9

[59] X. Qi, Y. Luo, G. Wu, K. Boriboonsomsin, and M. Barth, Deep reinforcement learning-based vehicle energy efficiency autonomous learning system, *Proc. IEEE Intelligent Vehicles Symposium, (IV)*, Redondo Beach, CA, June 11–14, 2017. DOI: 10.1109/ivs.2017.7995880. 7, 9

[60] Y. Hu, W. Li, K. Xu, T. Zahid, F. Qin, and C. Li, Energy management strategy for a hybrid electric vehicle based on deep reinforcement learning, *Applied Sciences*, vol. 8, no. 2, pp. 187, 2018. DOI: 10.3390/app8020187. 9

[61] Y. Li, H. He, J. Peng, and H. Zhang, Power management for a plug-in hybrid electric vehicle based on reinforcement learning with continuous state and action spaces, *Proc. 9th International Conference on Applied Energy, (ICAE)*, Cardiff, UK, August 21–24, 2017. DOI: 10.1016/j.egypro.2017.12.629. 9

[62] T. Liu, X. Hu, W. Hu, and Y. Zou, A heuristic planning reinforcement learning-based energy management for power-split plug-in hybrid electric vehicles, *IEEE Transactions on Industrial Informatics*, DOI: 10.1109/tii.2019.2903098. 7, 9

[63] P. Zhao, Y. Wang, N. Chang, Q. Zhu, and X. Lin, A deep reinforcement learning framework for optimizing fuel economy of hybrid electric vehicles, *Proc. 23rd Asia and South Pacific Design Automation Conference, (ASP-DAC)*, January 22–25, 2018. DOI: 10.1109/aspdac.2018.8297305. 7, 9

[64] R. Liessner, C. Schroer, A. Dietermann, and B. Baker, Deep reinforcement learning for advanced energy management of hybrid electric vehicles, *Proc. of the 10th International Conference on Agents and Artificial Intelligence, (ICAART)*, vol. 2, pp. 61–72, 2018. DOI: 10.5220/0006573000610072. 8

[65] Y. Hay, M. Kuang, and R. McGee, Trip-oriented energy management control strategy for plug-in hybrid electric vehicles, *IEEE Transactions on Control System Technology*, vol. 22, pp. 1323–1336, 2014. DOI: 10.1109/TCST.2013.2278684. 8

[66] E. Ozatay, S. Onori, J. Wollaeger, U. Ozguner, G. Rizzoni, D. Filev, and S. Di Cairano, Cloud-based velocity profile optimization for everyday driving: A dynamic-programming-based solution, *IEEE Transactions on Intelligent Transportation System*, vol. 15, pp. 2491–2505, 2014. DOI: 10.1109/tits.2014.2319812. 8

[67] H. Achour and A. Olabi, Driving cycle developments and their impacts on energy consumption of transportation, *Journal of Cleaner Production*, vol. 112, pp. 1778–1788, 2016. DOI: 10.1016/j.jclepro.2015.08.007. 13

[68] B. Jia, Z. Zuo, G. Tian, H. Feng, and A. P. Roskilly, Development and validation of a free-piston engine generator numerical model, *Energy Conversion and Management*, vol. 91, pp. 333–341, 2015. DOI: 10.1016/j.enconman.2014.11.054. 14

[69] J. Q. Li, Energy distribution and management of hybrid power system of electric drive vehicle (in Chinese), Doctoral dissertation in Beijing Institute of Technology, 2005. 14

[70] J. Xu, J. Ma, Q. Fan, S. Guo, and S. Dou, Recent progress in the design of advanced cathode materials and battery models for high-performance lithium-X (X = O2, S, Se, Te, I2, Br2) batteries, *Advanced Materials*, vol. 29, no. 28, 1606454, 2017. DOI: 10.1002/adma.201606454. 15

[71] A. Khalid, N. Javaid, M. Guizani, M. Alhussein, K. Aurangzeb, and M. Ilahi, Towards dynamic coordination among home appliances using multi-objective energy optimization for demand side management in smart buildings, *IEEE Access*, vol. 6, pp. 19509–19529, 2018. DOI: 10.1109/access.2018.2791546. 17

[72] S. B. Kotsiantis, I. Zaharakis, and P. Pintelas, Supervised machine learning: A review of classification techniques, *Emerging Artificial Intelligence Applications in Computer Engineering*, vol. 160, pp. 3–24, 2007. 18

[73] R. S. Sutton and A. G. Barto, *Reinforcement Learning: An Introduction*, MIT Press, 2018. DOI: 10.1109/tnn.1998.712192. 19

[74] J. Gläscher, N. Daw, P. Dayan, and P. O'Doherty, States v. rewards: Dissociable neural prediction error signals underlying model-based and model-free reinforcement learning, *Neuron*, vol. 66, no. 4, pp. 585–595, 2010. DOI: 10.1016/j.neuron.2010.04.016. 19

[75] T. W. Killian, S. Daulton, G. Konidaris, and F. Doshi-Velez, Robust and efficient transfer learning with hidden parameter Markov decision processes, *Advances in Neural Information Processing Systems*, pp. 6250–6261, 2017. 20

[76] Y. Bu, S. Zou, Y. Liang, and V. V. Veeravalli, Estimation of KL divergence: Optimal minimax rate, *IEEE Transactions on Information Theory*, vol. 64, no. 4, pp. 2648–2674, 2018. DOI: 10.1109/tit.2018.2805844. 21

[77] C. J. Watkins and P. Dayan, Q-learning, *Machine Learning*, vol. 8, no. 3–4, pp. 279-292, 1992. DOI: 10.1007/bf00992698. 23

[78] Y. H. Wang, T. H. S. Li, and C. J. Lin, Backward Q-learning: The combination of Sarsa algorithm and Q-learning, *Engineering Applications of Artificial Intelligence*, vol. 26, no. 9, pp. 2184–2193, 2013. DOI: 10.1016/j.engappai.2013.06.016. 23

[79] E. Rombokas, M. Malhotra, E. Theodorou, E. Todorov, and Y. Matsuoka, Reinforcement learning and synergistic control of the act hand, *IEEE/ASME Transactions on Mechatronics*, vol. 18, no. 2, pp. 569–577, 2013. DOI: 10.1109/tmech.2012.2219880. 23

[80] Y. Tseng, K. Hwang, W. C. Jiang, T. C. Huang, and S. S. Chen, An improved Dyna-Q algorithm based in reverse model learning, *ICSSE*, pp. 200–212, June 2015. 23

[81] M. Santos, V. López, and G. Botella, Dyna-H: A heuristic planning reinforcement learning algorithm applied to role-playing game strategy decision systems, *Knowledge-Based Systems*, vol. 32, pp. 28–36, 2012. DOI: 10.1016/j.knosys.2011.09.008. 24, 25

[82] T. Liu, X. Hu, W. Hu, and Y. Zou, A heuristic planning reinforcement learning-based energy management for power-split plug-in hybrid electric vehicles, *IEEE Transactions on Industrial Informatics*. DOI: 10.1109/tii.2019.2903098. 25

[83] I. Chadès, G. Chapron, M. J. Cros, F. Garcia, and R. Sabbadin, MDPtoolbox: A multi-platform toolbox to solve stochastic dynamic programming problems, *Ecography*, vol. 37, no. 4, pp. 916–920, 2014. DOI: 10.1111/ecog.00888. 25, 45

[84] D. Huang, H. Xie, H. Ma, and Q. Sun, Driving cycle prediction model based on bus route features, *Transportation Research Part D: Transport and Environment*, vol. 54, pp. 99–113, 2017. DOI: 10.1016/j.trd.2017.04.038. 27

[85] Y. Wang, W. Wang, Y. Zhao, L. Yang, and W. Chen, A fuzzy-logic power management strategy based on Markov random prediction for hybrid energy storage systems, *Energies*, vol. 9, no. 1, p. 25, 2016. DOI: 10.3390/en9010025. 27

[86] F. Bender, M. Kaszynski, and O. Sawodny, Drive cycle prediction and energy management optimization for hybrid hydraulic vehicles, *IEEE Transactions on Vehicular Technology*, vol. 62, no. 8, pp. 3581–3592, 2013. DOI: 10.1109/tvt.2013.2259645. 27

[87] Z. Chen, L. Li, B. Yan, C. Yang, C. Martínez, and D. Cao, Multimode energy management for plug-in hybrid electric buses based on driving cycles prediction, *IEEE Transactions on Intelligent Transportation System*, vol. 17, no. 10, pp. 2811–2821, 2016. DOI: 10.1109/tits.2016.2527244. 27

[88] F. Tianheng, Y. Lin, G. Qing, H. Yanqing, Y. Ting, and Y. Bin, A supervisory control strategy for plug-in hybrid electric vehicles based on energy demand prediction and route preview, *IEEE Transactions on Vehicular Technology*, vol. 64, no. 5, pp. 1691–1700, 2015. DOI: 10.1109/tvt.2014.2336378. 27

[89] C. Xiang, F. Ding, W. Wang, and W. He, Energy management of a dual-mode power-split hybrid electric vehicle based on velocity prediction and nonlinear model predictive control, *Applied Energy*, vol. 189, pp. 640–653, 2017. DOI: 10.1016/j.apenergy.2016.12.056. 27

[90] M. Zulkefli, J. Zheng, Z. Sun, and H. Liu, Hybrid powertrain optimization with trajectory prediction based on inter vehicle communication and vehicle infrastructure integration, *Transportation Research Part C: Emerging Technologies*, vol. 45, pp. 41–63, 2014. DOI: 10.1016/j.trc.2014.04.011. 27

[91] M. Debert, G. Yhamaillard, and G. Ketfi-herifellicaud, Predictive energy management for hybrid electric vehicles-prediction horizon and battery capacity sensitivity, *IFAC Proceedings Volumes*, vol. 43, no. 7, pp. 270–275, 2010. DOI: 10.3182/20100712-3-de-2013.00066. 27

[92] Z. Yao and W. L. Ruzzo, A regression-based K nearest neighbor algorithm for gene function prediction from heterogeneous data, *BMC Bioinformatics*, vol. 7, no. 1, p. S11, BioMed Central, March 2006. DOI: 10.1186/1471-2105-7-s1-s11. 28

[93] D. P. Filev and I. Kolmanovsky, Markov chain modeling approaches for on board applications, *Proc. IEEE American Control Conference*, pp. 4139–4145, 2010. DOI: 10.1109/acc.2010.5530610. 30

[94] D. P. Filevand and I. Kolmanovsky, Generalized Markov models for real-time modeling of continuous systems, *IEEE Transactions on Fuzzy Systems*, vol.22, pp. 983–998, 2014. DOI: 10.1109/tfuzz.2013.2279535. 30, 35

[95] Y. LeCun, Y. Bengio, and G. Hinton, Deep learning, *Nature*, vol. 521, no. 7553, p. 436, 2015. DOI: 10.1038/nature14539. 31

[96] S. Di Cairano, D. Bernardini, A. Bemporad, and I. V. Kolmanovsky, Stochastic MPC with learning for driver-predictive vehicle control and its application to HEV energy management, *IEEE Transactions on Control Systems Technology*, vol. 22, no. 3, pp. 1018–1031, 2013. DOI: 10.1109/tcst.2013.2272179. 35

[97] O. Sundstrom and L. Guzzella, A generic dynamic programming Matlab function, *Proc. IEEE Control Applications and Intelligent Control, (ISIC)*, pp. 1625–1630, 2009. DOI: 10.1109/cca.2009.5281131. 44

[98] X. Zeng and J. Wang, A parallel hybrid electric vehicle energy management strategy using stochastic model predictive control with road grade preview, I *EEE Transactions on Control Systems Technology*, vol. 23, no. 6, pp. 2416–2423, 2015. DOI: 10.1109/tcst.2015.2409235. 44

[99] J. Ko, S. Ko, H. Son, B. Yoo, J. Cheon, and H. Kim, Development of brake system and regenerative braking cooperative control algorithm for automatic-transmission-based hybrid electric vehicles, *IEEE Transactions on Vehicular Technology*, vol.64, no. 2, pp. 431–440, February 2015. DOI: 10.1109/tvt.2014.2325056. 46

[100] V. Mnih, A. Badia, M. Mirza, A. Graves, T. Lillicrap, T. Harley, and K. Kavukcuoglu, Asynchronous methods for deep reinforcement learning, *Proc. of the International Conference on Machine Learning*, pp. 1928–1937, June 2016. 76

Author's Biography

TENG LIU

Teng Liu received a B.S. degree in mathematics from Beijing Institute of Technology, Beijing, China, in 2011. He received his Ph.D. degree in automotive engineering from Beijing Institute of Technology (BIT), Beijing, in 2017. His Ph.D. dissertation, under the supervision of Prof. Fengchun Sun, was entitled "Reinforcement Learning-Based Energy Management for Hybrid Electric Vehicles." He worked as a research fellow in Vehicle Intelligence Pioneers Ltd. for one year. Now, he is a member of IEEE VTS, IEEE ITS, IEEE IES, IEEE TEC, and IEEE/CAA.

Dr. Liu is now a postdoctoral fellow at the Department of Mechanical and Mechatronics Engineering, University of Waterloo, Ontario N2L3G1, Canada. Dr. Liu has more than eight years' research and work experience in renewable vehicle and connected autonomous vehicle. His current research focuses on reinforcement learning (RL)-based energy management in hybrid electric vehicles, RL-based decision making for autonomous vehicles, and CPSS-based parallel driving. He has published over 40 SCI papers and 15 conference papers in these areas. He received the Merit Student of Beijing in 2011, the Teli Xu Scholarship (Highest Honor) of Beijing Institute of Technology in 2015, "Top 10" in 2018 IEEE VTS Motor Vehicle Challenge, and sole outstanding winner in 2018 ABB Intelligent Technology Competition. Dr. Liu is a workshop co-chair in the 2018 IEEE Intelligent Vehicles Symposium (IV 2018) and has been a reviewer in multiple SCI journals, including *IEEE Trans. Industrial Electronics, IEEE Trans. on Intelligent Vehicles, IEEE Trans. Intelligent Transportation Systems, IEEE Transactions on Systems, Man, and Cybernetics: Systems, IEEE Transactions on Industrial Informatics,* and *Advances in Mechanical Engineering.*

Printed in the United States
by Baker & Taylor Publisher Services